The Vision of Cosmic Order in the Vedas

The Vision of Cosmic Order in the Vedas

Jeanine Miller

Foreword by
Raimundo Panikkar

ROUTLEDGE & KEGAN PAUL
London, Boston, Melbourne and Henley

First published in 1985
by Routledge & Kegan Paul plc

14 Leicester Square, London WC2H 7PH, England

9 Park Street, Boston, Mass. 02108, USA

464 St Kilda Road, Melbourne,
Victoria 3004, Australia and

Broadway House, Newtown Road,
Henley on Thames, Oxon RG9 1EN, England

Set in Linotron Palatino
by Input Typesetting Ltd, London
and printed in Great Britain
by Billings Ltd, Worcester

Library of Congress Cataloging in Publication Data

Miller, Jeanine.

The vision of cosmic order in the Vedas.
Bibliography: p.
Includes index.
1. Vedas—Criticism, interpretation, etc.
2. Cosmology, Hindu. I. Title.
BL1112.26.M54 1985 294.5'921 84–24893

British Library CIP data also available

ISBN 0–7102–0369–1

TO I.T. AS EVER

TO DR PANIKKAR

Contents

Foreword

The Harmony of the Vedic World

The main significance of this thorough study, on a central theme of Vedic culture, and ultimately human wisdom, lies not – it seems to me – in its impeccable scholarship, sound ideas and wealth of information, but in the atmosphere in which the attentive reader is enwrapped and without which there is little hope of truly understanding any basic intuition of another culture. The context is as essential as the text. This is all the more important because, I submit, the Vedic Experience could play a vital role in alleviating the present day human situation. As a matter of fact western culture, spreading itself thin over the planet, seems to destroy itself, or fall prey of a death instinct. Yet, other cultures appear somewhat powerless to contribute to solving, or at least easing, the human predicament of contemporary Man.

By atmosphere I mean that imponderable context formed by the words used, the perspectives offered, the texts chosen and the entire *dhvani*, as the later indic culture would say, which puts the reader in the proper mood to reflect existentially on, and eventually incorporate into one's *sense of reality*, some of the insights expressed in this book. Let me explain.

This is not a book of hindu apologetics, nor one of those uncritical though well-intentioned writings proposing solutions to the world or indulging in a neo-hindu messianism of western flavour. It simply belongs to what I have called elsewhere the third stage of indological studies. In the first stage the main task was to collect, decipher, and interpret manuscripts in order to let the west know what India is all about. The main function of the second stage was to understand indic culture in its different aspects or as a whole. Indic culture ceases then to be a curiosity and slowly becomes a partner. The third stage, building on the previous ones, is formed by an attitude ready to learn and to be crossfertilized. This third stage is, among other things, critically conscious of two facts. First, each understanding of the other is done in parameters of our own self-understanding. We cannot do otherwise: we are conscious that we translate, interpret and, in a sense, transform (and even deforın) the indic sense of reality when expressing it in a foreign language. Second, the very effort of understanding the other changes our own self-understanding so that spontaneously a sort of mutual fecundation takes place. Trying to understand the other we better understand ourselves. We discover, for instance, some of our own presuppositions, which we took uncritically for granted, and change them according to the new insights. In the first stage the standards and methods of western scholarship are considered as sort of absolutes. The indic pantheon, for instance, is seen over and against the background of abrahamic monotheism or post-enlightenment rationalism. God means the abrahamic God, creation just production *ex nihilo* and once and for all, reason just *ratio* and its variances. The flaw of such a procedure appears immediately the moment we think of the sanskrit words we use for God, creation and reason – in these cases. In the second stage one tries to use and assimilate the indic categories. The indic *pramāṇas,* for instance are given their own due and a certain autonomy: *buddi, cit, manas, prajñā,* etc., are differentiated and tried to be understood on their own terms and not just univocally related to corresponding greek, latin or more modern words. In the third stage one realizes that the indic ways of life and thinking interpellate or even challenge the western paradigms, and that these latter may need to be changed in order to encompass without undue violence another different vision of reality. A certain transformation (*pariṇāma*) and even conversion (*metánoia*) seems to be required.

Jeanine Miller is one of the contemporary scholars today who applies more than just a methodological empathy in order to study the Vedas. A certain type of modern exegetes, east and west, *believe* that only a so-called impartial, unbiased and detached analysis will yield 'scientific' results. One can understand this attitude as a reaction against the uncritical claims of many 'believers' and as a result of the *belief* in a purely objective and universal reason untouched by space, time and culture. The most you can offer in that case is a condescending empathy in order to make some sense of the 'primitive', 'pagan', pre- or alogical ways of thinking of other people. Even if confused, 'galimatias', 'hocus-pocus', or exotic, other people's views are always 'interesting'.

This is not the place to discuss these momentous problems. It is only an occasion to show how a believing and yet not dogmatic approach can open us up to an understanding from within and yet not preventing a critical look from without. This is what Jeanine Miller tries to do. She thus explodes old views, like that of the naturalism of the Vedas, and leads us to understand the *sense of reality* of Vedic Man on a deeper level than that of the first two stages of indological studies. I use this phrase, 'sense of reality' and not worldview, philosophy, religion or the like in order not to unnecessarily superimpose overloaded concepts. It is not easy for contemporary Man to experience an undivided universe in which the *triloka* of the cosmic, the human and the divine or in other words the *adhibhautika*, the *adhyātmika*, and the *adhidaivika* have either not been set apart or not substantialized as independent entities. What if these three dimensions of reality, were three adjectives precisely of what the Vedas called *ṛta*, the greeks *eînai*, the buddhist *śûnyatā* and some moderns *reality*? But I should limit myself to present Jeanine Miller's detailed study.

Yet, the best way, it seems to me, is not just to praise what she has done or merely to repeat it on another register, but to reflect on the data she has gathered and the interpretation she has suggested inviting the reader to go a step farther in the same direction. In order to be brief I shall not quote a single text. The reader will find them in the book and I am leaning on them as well as on my own studies on the subject. One could equally call a postface what now follows. But the roots of the cosmic tree are in the air above and the branches below in our indic tradition. So a postface can equally be placed as preface.

First of all we may like to note three major topical absences in the exposition of the author. I would have preferred at least some mention of them. But living as she lives in the contemporary west and in contact with the French literature in the British Museum where she works, I can understand why consciously or unconsciously she may have had enough of them and has preferred to pass under silence those three great topics of contemporary western civilization: sex, psychology and politics, which obviously also have their role in vedic culture. I elaborate.

Any visitor to India, any expert in purānic lore, or any knower of the Vedas will certainly notice that sexual themes are not absent in indic culture. We have even specialized literature on the subject, not excluding the Vedas. In point of fact sex is not a vedic taboo. References are implicit and explicit. Jeanine Miller prefers to practically overlook all those motives. I surmise the reason why. Not only does she want to avoid sensationalism; she wants to obviate false interpretations. After all sex is only the last manifestation of a human, cosmic, and divine polarity and we cannot mention those creative tensions without entering into an anthropological field, which unless duly explicitated, would not help the understanding. The Cosmic Order shows a polar dynamism. To go all the way from there until human sexuality would require such an amount of provisos and distingos that the author has preferred to avoid them. I imagine another pedagogical reason also. The moment that we introduce sexual imagery and language, the atmosphere I spoke about in the beginning fades away for a good number of readers. It becomes too spicy and counter-effective to the idea of harmony she intends to convey. Patriarchy and matriarchy are important topics; the concept of *ṛta* may be precisely a hint of how to overcome or avoid such splits. But silence may be equally eloquent.

Another absentee in the book and related to the previous one is the psychological interpretation so common nowadays. I understand this relative absence on more positive grounds than the previous one. The reason is indological and cross-cultural. I am not personally against psychological interpretation, nor is the author as far as I know. I consider Jung, for instance, extremely useful for understanding many phenomena and no contemporary writer can today with impunity ignore Freud, just to quote two major, but not exclusive names. Yet, the danger of superimposing foreign categories in order to understand a fundamental intuition like that of *ṛta* is so great that Miss Miller's silence is a positive asset. All too often we claim to understand something better than the authors themselves or to interpret a culture in a deeper way than the 'natives' themselves. It is not impossible; but before this second step is undertaken we should first learn and try to experience the basic insights of an author or a culture in their own terms. I hinted in the beginning at the dangers of intellectual colonialism when mentioning the three stages of indological studies. The fact that psychological explanations are convincing 'for us' does not mean that they are valuable to explain a foreign culture.

This belongs already to the cross-cultural reason hinted at a moment ago. Whatever the merits and/or demerits of the psychological approach it is not an autochthonous explanation. Thus, lest we commit a sin of cultural reductionism we should apply modern western psychological categories only with great care and subsidiarily to the vedic sense of reality. Modern psychology, for instance, presupposes an individualistic consciousness hardly understandable in the vedic world. I am not saying that strong personalities, individual heros or indiosyncratic types are lacking in classical India. I am only affirming that the modern western individualistic perspective is foreign to vedic culture in which all is interconnected and the function of a bigger complex, as the famous upanishadic correlations, for instance, show.

The third silence is more delicate. My task is neither to defend nor to criticize but to introduce this book. No need to have a marxist reading of history and interpret the vedic phenomenon as the result of the political victory of the aryans over the dravidians, the protoindians or whoever the first inhabitants of the indic subcontinent were. No need either to interpret the caste system as sheer exploitation of the upper classes. Yet, the *Sitz im Leben* of the vedic civilization is precisely the fecund syncretic amalgam of two peoples and conceptions of reality. The Vedas are the result of an encounter, social, political, religious. The end product is a stupendous, probably new, sense of reality, a new culture, but the ingredients show the scars of a heavy struggle. Jeanine Miller does not tell us the gestation of the concept of *ṛta*, how this power intuition came into being, how power, prestige and domination may have played an important task in justifying a certain conception of cosmic order. But here again, I refer to the atmosphere she creates in her book by not entering into historical considerations. Important as they are they might distract us from the spiritual insight. She follows in this the most genuine indic tradition of ignoring historical gestation – even if it is retorted that this is to play the game of pandits and brahmins.

I have been mentioning all this not to correct the author indeed, but to situate this book in its most general framework without entering into minor details of underscoring so many happy findings.

I may try now to re-enact the fundamental experience of *ṛta* and to explain in plain words the sense of reality revealed in that word.

I may mention here the joy of finding a person with whom I can agree so much. We have struck such similar ideas thanks to an immediate contact with the texts and with the living tradition of India. It is similar to the joy of two scientists realizing that they have made the same discovery independently of each other.

I have mentioned the atmosphere of this book and three

absent major topics. Let me put an example to introduce the experience of *ṛta* and to link it with what I have been saying.

The author, of course, speaks of *soma*, the powerful vedic drink, and yet she omits the topic – so hotly debated in other circles – as to whether it is an intoxicating drug and whether we can identify it chemically. Here we have a clear case of the insidious biases of questions. Under the pretext that questions let us free to give any answer, they condition already our ways of thinking and impose on us a particular perspective, namely that for which the question is relevant. There is no doubt that there is a powerful drink called *soma*, as there is no dispute regarding the christian fact that the eucharist has something to do with real bread. But the important eucharistic problems lie elsewhere and do not include that of the transsubstantiation which is already a distraction of the real question. The real questions concern the presence of Christ among us, the divinization of the Christian, the heart of the Christian community and the like. Similarly, whether *soma* is a drug or not and what kind of effects it has distracts us from the real question of a *soma*-spirituality and the sense of reality connected with it and with our *ṛta*. There may be – there are – psychological experiences at the bottom, but the vedic question is not about that. It is about an integral experience which mirrors reality and which even makes it.

The experience of *ṛta* is not primarily a psychological experience. It is one of the most powerful symbols of the vedic tradition and probably ultimately discloses to us the vedic myth, what some today would call the cosmological myth.

Let me put a parallelism of what I have called elsewhere an homeomorphic equivalent, i.e. that concept or set of concepts which play the corresponding equivalent role in another cultural system. The homeomorphic equivalent in this case is *evolutionary history*. This plays a partially equivalent function to the symbol of *ṛta* for vedic Man. For the average modern western Man evolutionary history is the womb (*yoni*) where reality unfolds; it is what holds all things together (*dharma*) and what makes us understand and see (*dhīḥ*). Within the western myth of history something is real when it is an historical fact, something is intelligible when we can discover its historical genesis, how it has evolved until the present situation. Evolutionary history, in this sense, is both the manifestation of reality and that out of which the manifestation comes. The modern world is an historical world and evolution is the key to intelligibility. Reality is history and history the criterion for reality. They are not identical: there can be dreams and projections, thoughts and impulses; but as long as they are not historical crystallizations or historically detectable, they are not considered truly real. History is the background over and against which modern Man experiences reality. Yet history is not primarily a psychological experience, induced by the modern 'drug' of 'education'.

Something equivalent could be said about *ṛta*. It is the warp and woof of reality which is being woven in the spinning loom of the sacrifice. *Ṛta* manifests the cosmotheandric activity of reality itself, i.e. the dynamic interaction between the cosmic, the human and the divine factors of the real. It is the very expression of the harmony of the All (*sarvam*), of all that is.

Ṛta is translated by our author as harmony and quite rightly so. It is what links all together, the link itself and the ultimate dynamism of the real. There is but one word which has slipped the pen of Jeanine Miller which I would like to qualify. In order to make the western and westernized (english speaking) reader understand she adds that *ṛta* represents the Law of harmony. And rightly so in as far as *ṛta* is not chaos. It is order, in a way similar to the *ordo* of the christian scholastics which combines the greek *taxis* and the intelligent, creative, and provident power of the patristic-medieval God. Whatever this may be, perhaps the word 'law' conveys too much. Science can be elaborated from observing the laws of nature. *Ṛta* does not allow any such scrutiny. It is sovereignly free. It is not constricted by any law. Here we touch a topic dear to me. The western polarity, which since Parmenides reigns in the western world, does not rule in the indic culture. The ultimate pair in the west since Greece is that between Thinking and Being. The former can theoretically discover the laws of Being. Thinking is paramount. Being is tamed, subservient. Therefore, contra*diction* is also not valid for Being and not only for Thinking. *Ṛta* has none of these strictures. There is no law of *ṛta*. There is *ṛta* and *ṛta* is harmony; but this harmony is not subjected to any ulterior law. There is no mind behind.

To live in a *ṛtic* universe represents a fundamental human experience different from that of believing to live in a logical world or in a universe, governed by a law. No need of a supreme God to warrant the order, no need of strict 'scientific' thinking in order to survive. Thinking, Science, Knowledge, all are important but they do not have an absolute status. This is what the Upănișads will try to qualify later. Being is free, ultimately even from thinking. No need of absolute ethical norms at the ultimate level. No need of fear, 'Angst', anxiety, regarding ultimate questions. *Ṛta* is there, but not as a refuge. No need to control everything, to be certain of all things, to know everything. No need to live facing the future or scrutinize frantically the past. We may trust in *ṛta*. It is not for me – and much less in a preface – to spell out how contemporary Man can learn from this basic sense of reality. An attentive reading of this book may help.

I have been just saying that this book demands to be read as we enjoy poetry. Poetry is truly such when it is *apauruṣeya*, i.e. without a human author, because once we are led to utter in truth the words we read we discover ourselves being the author. We are ourselves the artificers (*poiētai*) of what is said in the poem. Or, in the beautiful intuition of the author – did she need to be a woman? – when she describes to us that the crux of the matter is the heart.

R. Panikkar
Tavertet
Autumn 1984

INTRODUCTION

The hymns of the Vedas, the four collections of which the Ṛgveda is the oldest and the basis of the three others, which western scholarship dates back to about 1500 to 1200 B.C., form the original source book of Indian religion. Revered till today, the Ṛgveda is at present little known in India. Western scholars, headed by Max Müller, could find in it not much more than the expression of primitive man's wonder at nature. But far more than this, the Vedas are the legacy of a whole civilization, a mirror into man's psyche and its vision of life in all its intricacies. Undeniably, the Vedic hymns are deeply imbued with the poetry of and feeling for nature, and this constitutes part of their charm. But behind the mythological vision of the cosmos emerges a remarkable insight into the workings of the universe, the underlying oneness of all life behind the multiplicity of forms and forces that hide it, the one law of transformation, of growth and decay, of evolution and involution, of order and adjustment whereby all things are kept more or less in equilibrium, all disorders being restored back to order, this causing friction, conflict, and in man's case, suffering.

The great archetypal symbols of Heaven and Earth as parents, of sun and moon as fosterers, of bull and cow as impregnators, of water and fire as cleansing and creative agents, of tree and bird or swan as image of the cosmos or the human psyche, the great archetypal ideas such as 'cosmic order' (*ṛta*), or 'sacrifice', or 'truth' or 'duty', are all not only outlined, but elaborated, and form the basis of that remarkable development of philosophy which took place on Indian soil ever since the composition of the Vedas.

The present book, based upon the author's thesis for the

degree of M. Litt, taken in 1979, investigates the Vedic sages' vision of the Cosmic Order as one can reconstruct it from the scattered hints and references. Three fundamental aspects of the Law of Cosmic Order are studied: 1) the law as it manifests in the universe and therefore nature, the cosmic aspect; 2) the socio-ethical aspect which, in the human context, can be translated as integrity-integration, truth at the human level being equivalent to harmony at the universal level; 3) the religio-sacrificial level which investigates the little understood meaning of the one sacrifice with which the cosmic order is identified – not only in the Vedic conception but in many of the religious traditions of past ages. This vision of cosmic wholeness was never surpassed by any subsequent religious philosophy, and has not been investigated as a whole.

The Sanskrit word *ṛta*, which we have rendered as *Cosmic Order*, admits of a wide range of meanings, according to contexts: the cosmic harmony, the truth, the right, or righteousness, the ritual or rite that expresses the Order. The Vedic conception of the cosmic order, the dynamics of manifestation, *ṛta*, was developed from a vision reaching out to vast dimensions and arrived at by generations of seers. The vision itself embraces a wide domain of activities, cosmic, natural and human, showing insights into the laws governing the universe, the gods' and men's role therein. These insights range from the ṛṣis' speculation as to the origin of creation, the orderly unfoldment of the universe and its various kingdoms in accordance with strict laws, the function of the gods, with particular stress upon the mutual interaction of the godly and human realms – the socio-ethical order worked out for humanity being a direct reflection of the universal law governing and governed by the gods – to the sacrificial ritual as man's microscopic re-enactment of the cosmic order envisaged as a stupendous rite. The whole process of cosmic unfoldment that emerges from this Vedic vision appears as a universal offering of all the myriad lives to the One Life, the Puruṣa, that underlies and sustains them all.

The hymns of the Vedas, the songs which the ṛṣis composed as expressions of their vision, were subsequently called *śruti*, that which has been heard: *revelation* – the revelation granted to the sages and handed down from generations to generations by word of mouth.

The Vedas, in their essence, have been claimed to be a timeless revelation, the transcendent made manifest at the empirical level by means of visions couched into words through the poetic inspiration of illumined sages. Vedic tradition refers to the origin of the Vedas as *apauruṣeva*, non-human, i.e. its inspiration belongs to the eternal, it is a manifestation of the word, *vāc*, the primeval vibration at the basis of all things and dwelling in the heart of the ṛṣis.

This implies certain issues, of which one – the fact of 'vision' – has been investigated by J. Gonda,[1] but of which the most important one, what exactly is that vision, that revelation, has been mostly ignored by Western exegesis. A. C. Bouquet, in his study, *Hinduism*, voices a doubt, as scholars and students of the Vedas have all entertained, as to what exactly those Vedic hymns reveal: 'Since, however, most of them consist of lyrical addresses to the greater deities of the people concerned, it is difficult to see what, in a literal sense, they reveal.'[2]

This is the problem and one that has militated against the understanding of the Vedas as revealed scriptures. The revelation is certainly there, but couched in such a way that it is not quite apparent, at least to the Western mind. As the hymns were songs of praise or invocations to the many gods, they were not intended to give any explanation; whatever we can glean from them is by the way, the whole vision of the sages is, so to speak, hidden behind praises; it is the foundation structure which is taken for granted and only occasionally alluded to here without any attempt at systematising the basic knowledge derived from the vision or revelation.

Confronted with a multiplicity of gods, of praises and invocations and sacrifices, of obscure allusions, riddles and myths, Western exegesis completely failed to perceive the vision of Oneness, of godly and human interlinkedness, of divine solidarity, of cosmic harmony which underlies this bewildering array of activities, images and names, the vision of cosmic integrity which is the Ṛgveda's peculiar contribution to religious speculation and which it is the purpose of this book to unravel. Indeed, scholars followed the trends of the various schools of thought prevalent in their day, all based upon the evolutionary theory whereby prehistoric man could only express primitive, i.e. simplistic thought; and the many gods could never be admitted to be the many manifested expressions of an underlying unmanifest yet ever present Oneness, in spite of remarks scattered here and there in the Ṛgveda:

> One whole governs the moving and the stable, that which walks and flies, this variegated creation (III.54.8).
>
> That which is one has developed into the all (VIII.58.2d).

The Vedas were judged at their face value, enquiries never went much beyond the surface meaning. In this respect, it is interesting to note the completely different assessment of the Vedas given by the nineteenth century Indian sage, Ramakriṣṇa: for him, the last attempt at expressing man's inkling of the Absolute is found in the Vedas, a conclusion to which no Western scholar would subscribe.

But let us consider what is the foundation of Ramakriṣṇa's statement. The conception of the Absolute (that One, *tad ekam*) cannot really be defined in words, for any word or thought by definition limits the very absoluteness of the idea that is attempted to be expressed. The ṛṣis of Ṛgvedic times nevertheless tried to express their vision, revelation, illumination, in three ways:

First, negatively, by eliminating everything that can be known to the mind of man and posited concerning Deity. Such a way may be considered a reduction to nothingness of that which is supremely all-inclusive and yet beyond all inclusiveness; but the result, as voiced by inspired poets, gives rather an impression of sublime simplicity and extraordinary heights scaled by the human mind, those heights themselves being but a poor approximation, a completely distorted mirror, of what indeed is reality. All mystics of subsequent ages in their attempts at expressing Deity have followed more or less this approach.

Second, by showing the Absolute as that which is beyond the sway of the opposites which rule this world of manifestation, beyond night and day, i.e. darkness and light and all that this implies, hence 'good' and 'evil', as that

> whose shadow is death, whose shadow is immortality (X.121.2)

and yet that whose essence inheres to the whole of cosmos through the latter's manifold agents, the *devas* or shining ones, usually translated as 'gods', and whose very breath is the eternal movement or rhythm that animates the whole.

Third, through the panorama of *cosmic harmony,* the interplay of the many, their interlinkedness, the manifold variety of the cosmos being as it were the mirror of the manifold possibilities inherent in the One out of which proceeds 'this all'. This panorama forms the woof of the Vedas.

Furthermore, in the words of R. Panikkar: 'The Vedic Revelation . . . conveys a transtemporal message, if we are permitted to use this word with reference to the present, in order to make us more aware of the mystery of existence and of life.'[3]

To this vision may be traced the great themes of the later speculations of Indian philosophy. The Upaniṣads' negative approach to the Absolute is a direct descendant of the Ṛgvedic poem of creation. This Vedic vision is also the root of the Indian conviction that any philosophical doctrine, to stand the test of truth, must be based upon experience. The transcendent must be brought into the field of the existent by direct intuition or visioning, in order to be established as truth and communicated to others, however inadequately.[4]

To express such supra-sensuous knowledge, such revelation, is a task that only a body of inspired seers can undertake, and this only in moments of high tension, of lofty inspiration, of illumination. The resulting religious philosophy is thus rooted in the personal experience of not just one seer but generations of seers through the discipline of contemplative meditation.

The vision, such as it has come down to us, is scattered through the *Ṛgveda saṃhitā*, the oldest of the four collections called the Vedas, and has to be reconstructed piece by piece from the various hints, obscure allusions and myths. Many a conclusion one is led to draw is derived more from implied than explicit statements. The reconstruction of this vision, the examination of its insights, the conclusions that can be drawn in the field of philosophical thought and of socio-ethical and ritualistic practices, falls within the scope of the present study. We have concentrated mainly upon the text of the Ṛgveda, being the oldest and the basis of the Vedas.[5] Quotations from the Atharvaveda and the Yajurveda have occasionally been used to substantiate any particular point; quotations from the Brāhmaṇas and Upaniṣads have also been cited to show the similarity in or development of thought. As to the question of interpretation, the traditional idea that the Veda[6] bears a threefold interpretation *ādhyātmika* (spiritual), *ādhidaivika* (relating to the gods) and *ādhiyajñika* (ritual) has been favoured, the underlying aim being to examine the ṛṣis' claim to divine revelation and their hymns the expression of their inspiration. As the question of seership underlies the 'vision' as such, it will be examined in the first chapter.

Part I

VISIONING AND REVELATION

ONE

The means of knowledge

The means of valid knowledge

The ways of apprehending truth have always been considered of fundamental importance as a guideline to the evaluation of that very truth, for that truth can be blurred or distorted in accordance with the greater or lesser validity of the means. What is accepted as a valid mean?

Indian philosophical schools recognise at least three means (*pramāṇas*) to valid knowledge (*pramā*):

perception (*pratyakṣa*)
inference (*anumāna*)
verbal testimony (*śabda*) or revelation (*śruti*).

The testimony of the Vedas is concerned with the last of these, namely *śruti*. Some Indian logicians reject verbal testimony (*śabda*) as a means of knowledge. According to them verbal testimony can only be accepted if based upon the first two methods of apprehension – namely perception and inference which form the basis of daily experience. But knowledge, it was found, cannot be restricted merely to the senses, or direct perception and reason, as the only source we have at our disposal. For these do not exhaust the whole content of human experience. Therefore some schools of Indian philosophy recognised intuition or spiritual insight as a means of knowledge, a unique *pramāṇa*, differing from the others, yet valid in itself. The Veda, based upon extraordinary means of knowledge, in fact, upon the intuitive vision of seers, was put forward as such an acceptable source of knowledge. In due course of time its systematic interpretation became the aim of the Mīmāmsā system of philosophy.

The ṛṣis' means of apprehension are twofold: *visioning* and *contemplating*. Visioning may mean receiving knowledge as in a flash and thereby intuitively knowing, or it may mean 'seeing' certain realities beyond the world of the senses, both means implying direct apprehension of certain truths that do not fall within the field of ordinary human sensuous experience. Contemplating implies meditation, the steady focusing of the inner insight upon a particular object, or idea or vision. Intellectual inferences and logical deductions do play their parts, but they are secondary and only means of explaining and expanding upon what has been seen or discovered in a supramental state of consciousness.

The acceptance of revelation, *śruti,* as an authority was considered to have its dangers: anything could be believed as revelation. To obviate this danger, conditions were laid down: thus, revealed truth that refers to the extra-empirical must do so in terms of what is known, otherwise it would be unintelligible; nor should it contradict the known; it must be coherent, it cannot be against reason, though it may stand above it. The danger of complete dependence upon the authority of individual insight which, however superficially interesting, is liable to illusion or may be defective either through an unbalanced assessment of reality or in point of depth of insight, may be offset by the fact of plurality. The Veda's authority stems not just from the testimony of one ṛṣi, but of generations of seers. The testimony can be traced to a long line of sages.

On the other hand, the end-product may not be fully coherent, or may vary in the quality and depth of insight evidenced. Scholars usually bring forward the incompatibility of the cosmogonic theories of creation described in the Ṛgveda. There is indeed no attempt at systematising the insights received by means of inspiration, but the diversity of answers given, e.g. to the riddle of the origin of the universe, which for Western exegesis is a sign of conflicting or contradictory ideas, does not detract from the coherence of the whole; for it becomes obvious that the ṛṣis considered there could be no one definite, specific answer at our level of cerebral understanding to the great question of 'how' and 'why' the universe came about, but various answers had equal value in their eyes in as much as they showed the various facets of the problem, and that some of the answers given apply to different levels of manifestation, levels which Western exegesis completely failed to note.

To conceive of the beginning as a 'sacrifice' or sacrificial offering of some Supreme Being may seem poles apart from the idea of 'contemplation' as the supreme act of creation,[1] or from the idea of 'projection' from within without. It may seem poles apart from the idea of the eternal, all-powerful *word* or *sound* as the original creative agency and moulder of all;[2] or from the idea of the fashioning of all things out of primeval substance, whether called the 'waters' of space as is generally the case, or simply 'wood',[3] words which should not be taken literally. Lastly, it may seem poles apart from the idea of the rending asunder of a primeval monster typifying chaos. But this is exactly where one has first to take into account – not in a superficial way but in a profound analysis – the meaning of certain keywords and key-thoughts of the ṛṣis such as

tapas – the creative flame of contemplative exertion, the contracting to an innermost point of dissolution and the subsequent expansion to an infinitude of creative possibilities

yajña – the sacrifice, the worship, the setting up of a communication between the within and the without and thereby the making 'holy', so that the expanding without may be worthy of the within, may become a perfect mirror; hence the sacrifice in establishing a circuit of give and take, a balance, is the harmony

visṛṣṭi – the projection of the inner into the outer, the ultimate rational way of expressing the former steps

before one realises that none of these is mutually exclusive of the other; and secondly to appreciate that all of these expressions represent the one and same overall idea of creation in its many faceted aspects as it is worked out in different stages or levels of manifestation in a distinctively conceived unfoldment of the universe.

The contemplative exertion whereby the divine creative flame is released (*tapas*) – this in itself being the sacrificial offering of Deity (*yajña*), the giving up of blissful homogeneity so that a world of heterogeneity may be – is itself the primordial step, the origin of all. The One thereby becomes two: the Absolute beyond conception shrouded in darkness and remaining in darkness even in the period of manifestation, and the Creative Deity (the demiurge) born of the fire of divine contemplation (*tapas*), personified in Agni, in Varuṇa, in Viśvakarman and in Prajāpati. Here may be glimpsed the original idea of the unfoldment from within without, the first step, hence the 'projection'. The darkness that veils the ultimate secret is that primordial substance or 'waters' of space, that at a more differentiated level, yet still transcendental to our earthly realm, becomes the original basis of matter out of which the demiurge fashions the universe.

The colourful language of myth expressed all this as the dragon of chaos (more and more anthropomorphised as time went on) being rent asunder, dismembered, by the demiurge – in this case personified as Indra – as a result of which the waters of life are let loose, the sun is set on high, the darkness of chaos is changed into a world with divisions, spheres, realms, numbers, a world of light, symbol of consciousness emerging from the unconscious. At the terrestrial level, creation is repeated in a similar way, but expressed differently; here the great personification of nature's creativity, the god Tvaṣṭṛ, is depicted as presiding over all births and shaping all forms.

Each step in the manifestation of the world does not exclude or contradict the previous one, but all steps are expressive of the manifold ways of creation as perceived by the vision of the ṛṣis. Viewed in this way, the various attempts at defining the successive acts of creation dovetail each other and form a whole that constitutes the substratum of the Vedic vision of creation.

The validity of the Ṛgveda's claim to revelation

The ṛṣis are inspired men; they are called 'poets', *kavi*, in the ancient sense of the word. A *kavi* is not a juggler with words as our modern poets have all too often become, but a man inflamed, a man kindled with divine inspiration, 'flame-tongued', 'sun-eyed' like the gods; a seer with golden tongue. He gives voice to divine utterances whose truth lies in the illumination of which he is the recipient and the revealer.

Hints scattered throughout the Ṛgveda enable us to build up a substantial picture as to the seers' visionary ability and the conclusions they reached, though neither the one nor the others are fully described. Here again a deeper understanding of keywords is necessary. One of these, of fundamental importance to our present study, is *dhīḥ*. In the light of Professor Jan Gonda's examination of the Vedic word *dhīḥ* and its derivative *dhītiḥ* in *The Vision of the Vedic Poets*, it has become evident that *dhīḥ* is not just 'thought' as it is usually translated, or 'hymn' or 'devotion', but 'visionary insight', 'thought-provoking vision', a spiritual power which enabled the poets to see in depth and which they duly expressed as songs, hymns, or prayers, so that the word has often the meaning of insight-leading-to-song, hymn, or prayer. It could be likened to inspiration bursting forth into song.

J. Gonda's explanation of the meaning of *dhīḥ* and his reason for choosing the word 'vision' to translate it in most cases, is thus set forth:

> The Skt. noun *dhīḥ* is like vision closely associated with a verb expressing the idea of 'seeing'. By 'vision' is . . . to be understood the exceptional and supranormal faculty, proper to 'seers', of 'seeing', in the mind, things, causes, connections as they really are, the faculty of acquiring a sudden knowledge of the truth of the

> functions and influence of the divine powers, of man's relation to them, etc. It is this vision which they attempt to give shape, to put into words.[4]

Dhīḥ allies both seeing and intuiting; it is both a flashing forth of some truth and the expressing thereof in poetic utterance.

Thought, on the other hand, is generally based on data furnished by the senses and is an instrument of ordering what becomes accessible to the mind via the senses. It is used to put together a rational picture of that reality presented to the mind by sensory perception. But it may also be used to order that which has been received through intuition or spiritual insight, that which has been visioned at a level of experience quite divorced from the physical senses. In that case, the process of thought is first by-passed whilst a channel of vision is opened up and used to reach out beyond the senses and the mind, to those realms described in the Ṛgveda as 'beyond' (*para*). Once the vision has been experienced the thought process takes up its task of expressing it in terms which are understandable and acceptable to the mind, in other words, to make some sense out of what has been received.

The vision reveals. It may be that in Vedic times the word *dhīḥ* did also include the meaning of 'revelation', subsequently to be forgotten or dropped and kept only in the word *śruti*; insight and vision into the beyond are 'revealing' of certain factors beyond the sensory world. The bards are the revealers and interpreters of what they behold through their illumined vision. It may be of interest to compare the bards of Vedic India with the prophets of Israel, a comparison which has not been made (to the writer's knowledge). The word 'prophet' (*pro* forth, *phetes* speaker) refers to 'one who speaks for God or for any deity, as the inspired revealer or interpreter of his will' (*Oxford English Dictionary*). This definition fits the Vedic bards perfectly. They were the divine spokesmen and what we get in the Vedas are their songs, the by-products of their seership. The word 'prophet' has come to have the connotation of predicting the future which originally was secondary to 'inspired' utterer and interpreter. A good instance of its fundamental meaning is found in the Gospels in the episode of the woman of Samaria whom Christ asks for a drink by a well. When she realises that he reads through her, has 'seen' that she had five husbands in the past and is now living with a man who is not her husband, she exclaims: 'Sir, I perceive that thou art a prophet' (John 4:19). Christ has not yet made any prediction to her, he has only read her background. In Deuteronomy 13:1 the prophet is described as a 'dreamer of dreams'. Dreams and visions are the ways in which revelation came to the prophets. Through these the 'Lord', speaking in the first person, multiplies his communications:

> I have multiplied visions and have used similitudes by the ministry of the prophets (Hosea 12:10).

H. L. Ellison in *Men Spake from God* claims that 'the popular conception of the prophet as primarily a foreteller is alien to the thought of the Bible. . . . Ultimately it is the spiritual quality of his message which shows whether a man is a prophet or not.'[5] In Samuel we have the following explanation:

> Beforetime in Israel, when a man went to enquire of God, thus he spake, Come and let us go to the seer; for he that is now called a Prophet was beforetime called a seer (1 Sam. 9:9).

The prophet is thus primarily a 'seer'. Isaiah refers to 'the vision of Isaiah, the son of Amoz, *which he saw*' (Isa. 1:1); and in Daniel: 'I Daniel, had seen the vision and sought for the meaning' (Dan. 8:15). Moses is considered the prophet *par excellence*: he communed with Deity, he knew the Lord 'face to face' (Deut. 18:15; 34:10), he received the Lord's messages. Enoch (Jude 14), Abraham (Gen. 20:7, 17), and the Patriarchs (Ps. 105:15) are referred to as prophets in the Bible. They are spokesmen of God. According to H. L. Ellison, 'the source of the message is supernatural, not natural. It is derived neither from observation nor intellectual thought, but from admission to the Council Chamber of God.'[6] He cites Amos:

> Surely the Lord God will do nothing, but He revealeth his secret unto his servants the prophets (3:7).

A similar claim is made by the Vedic bards whose insight takes them 'consciously' (*manasā*) to the gods' 'Council' where they perceived (*apaśyam* I 'saw') 'celestial entities' (III.38.6), and through 'setting their mind to contemplation' they 'discover the primeval sacrificial formula' that can be used as a pathway to the gods (X.183). With 'mental eye' they 'see those who originally performed the sacrificial rite' (X.130.6cd). By insights gained in secret within their heart they approach the 'thousand-branched' – tree of life? – and enter the kingdom of Yama (VII.33.9), Yama presiding over those who have passed beyond the gates of our space-time world. Plunged in their soul-searching meditation, delving into their 'heart's wisdom' they discover 'the link between the created and the Uncreate' (X.129.4). Through their intuitive perception (*manīṣā*) they 'observe' Indra hastening to the call of the Poet.[7]

> They 'behold' the 'untiring herdsman, approaching and withdrawing by the pathways' (X.177.3).

These are all hints as to the quality of visionary insight and revelation received by the ancient ṛṣis.

The ṛṣis of India are known as such because they received visions, they could 'see' in a way which ordinary people cannot, and they could sing forth their vision. They 'saw' and 'heard' the Veda, the divine wisdom. The prerogative of the ṛṣi is to enter into communion with his chosen deity in his visionary moments, as much as any prophet of Israel with his Lord God. Only, the Vedic gods do not constantly give orders, especially to destroy. Vasiṣṭha, brooding upon the great Lord of Cosmic Order, beholds the God's form coming into focus, but lo! it appears not as he had imagined it to be, but as the countenance of Agni the flame divine:

> When just now going into the presence of Varuṇa, I contemplated the face of fire (VII.88.2).

Like Moses to whom

> the angel of the Lord appeared . . . in a flame of fire out of the midst of a bush . . . and God called unto him out of the midst of the bush. (Ex. 3:2).

The Vedic bard gains admission into the presence of the great God Varuṇa. No wonder that at other times when this divine presence is denied him, he cries out in his despondency

> What has become of our friendship of old when we two kept unhindered company?
> O Varuṇa, law-abiding, I did enter thy mighty mansion, thy thousand-gated home (VII.88.5).

The quality of the vision and of the visionary activity of the ṛṣis is markedly different from that of the prophets of Israel. These seem to have been mainly the mouthpiece of a fully anthropomorphic God's wrath against the iniquity of men, hence their constant thundering for righteousness and their oracles of doom. Not so the bards of the Ṛgveda. The gods of the Vedas are not yet fully anthropomorphised; they are more energy centres beneficent to mankind, intelligent principles, 'brothers in the same mother's womb' (VII.83.8). For the seers, to infringe the statutes of the gods that actualise on earth the Great Law is to 'injure as it were a friend' (X.89.8). The gods' most intimate representant and spokesman to the heart of the ṛṣis, their link between the celestial order and the human kingdom, is Agni, messenger of heaven and earth, the flame that burns within all things and all creatures and in the human heart, the deathless flame that reminds us of Deuteronomy:

> For the Lord thy God is a consuming fire (4:24).

The Vedic vision is more cosmic in the sense of being in oneness with nature, the heavens, the earth, the infinitude of space, the elements, the light of sun or dawn, the trees, plants, animals, in contrast to the vision of the prophets of Israel concerned far more with their people and their wrong reaction to their immediate problems, with human failing and human restoration. There is no 'foretelling' nor doom forecasting in the Vedas, nor can anything be found as horrifying and destructive as the command of Ezekiel's Lord God

> go ye after him through the city, and smite; let not your eye spare, neither have ye pity;
> Slay utterly old and young, both maids, and little children, and women; but come not near any man upon whom is the mark (Ez. 9:5–6).

which immediately lowers the value of the message of Ezekiel in the eyes of those who value the message of love of another prophet, Christ. With the Vedas there is no vision of gloom, but one of exaltation, vigour, might, *joie de vivre*. Only one verse in the Ṛgveda, concerning Soma's act of judgment, might be considered as coming a little closer to the Prophets' constant anathemas:

> The Knower, beholding all worlds, hurls the hated and irreligious into the abyss (IX.13.8).

This does not necessarily mean the abyss of complete annihilation, but the powerful image could be compared with Amos:

> Behold, the eyes of the Lord God are upon the sinful kingdom, and I will destroy it from off the face of the earth (Amos 9:8).

Wrath and revenge so prominent in the prophetic 'revelation' are hardly existent among the Vedic bards, with a few exceptions with regard to wrath. Prayers are addressed to Rudra, the most ambivalent of the gods, to avert his 'wrath', for his terrible shafts kill both men and creatures, and similarly to Varuṇa. But fear and deprecation are far from being the current attitude of Vedic man to his gods, nor can the gods' 'message' be one of doom when one considers the daily expectation of the bards to receive 'blessings' from their prayers, hence from their invocations to the gods, blessings not only in material but also in mental and spiritual gifts.

The relationship between the prophets of Israel and their God and the Vedic bards and their gods is thus totally different, but the one similarity lies specifically in the communion established between the human and the divine, the receiving of visions, the inspiration that bursts out in eloquence. The occasional mood of repentance and questioning, as between the seer Vasiṣṭha and his chosen deity Varuṇa, does not exemplify the common attitude of Vedic man to deity, which is one of joy, of coaxing, of adulation, of invocation and evocation. From their songs one can reconstruct the vision of cosmic harmony as the substructure of the ṛṣis' thought. From the prophets of Israel one receives the hope of resurrection, of a future of righteousness for men on earth and therefore of reconciliation between God and man (Isaiah 20:7–9 and Micah 4:1–5). Reconciliation is man's step towards harmony, that harmony which he finds in the heavens, that harmony which the Vedic bard found mirrored in his very heart and to which he gave his eloquence. In the Vedas there is no promise of a new world, because this world here and now is the playground of a multitude of experiences that forge man's soul; it constitutes the path that leads to the infinite. With Ezekiel, however, where the world for the Jews is already one of doom, the promise is given of a new temple, the vision of a cosmic temple arising out of the ruins of destruction, of exile, the promise of a divine Presence even amidst the exiled ones, because of the renewal of the heart:

> Make you a new heart and a new spirit. . . . I have no pleasure in the death of him that dieth . . . wherefore turn yourselves and live ye (Ez. 18:31–2).

Similarly with another prophet:

> For the earth shall be filled with the knowledge of the glory of the Lord as the waters cover the sea (Habakkuk 2:14).

But the Vedic bards have also a vision of a goal from beyond this playground of opposing forces:

> Seven landmarks have the wise established. For one of these may the troubled mortal set out. Thence liberated from the paths verily a pillar of life, he stands on firm foundations in the abode of the Most High (X.5.6).

Can we read in these lines, as it were a hint, that having

completed his pilgrimage on earth through repeated existences (cf.IV.54.2), the weary mortal now liberated from the paths becomes a pillar of life in the highest heaven? This stanza could be fruitfully compared with a verse of Revelation:

> Him that overcometh will I make a pillar in the temple of my God, and he shall go no more out: and I will write upon him the name of my God, and the name of the city of my God, which is new Jerusalem (Rev. 3.12).

What did John mean by 'overcometh' and 'go no more out'? Both the Ṛgveda and Revelation point to a pilgrimage the goal of which is the abode of the Most High whence there can be no more going out.

The inspiration of the Bible and, as a corollary, its 'unique' revelation, has for too long been considered the only truly 'divine' inspiration among the scriptures of the world and thereby biased any appreciation of other 'revelations'. There could be no true inspiration from other sources save that stemming from the 'one Living and Only God' with a specific name. Although specific names die with the specific races that create them, this was not recognised; the latest comers on the scene of this earth were the only ones to possess the whole truth. But more subtle than this is the prejudice with regard to the idea of God as specifically anthropomorphic, leading to *a priori* rejections of any other conceptions of Deity. This is evidenced in the lack of understanding of the Vedic idea of God among scholars of the turn of the century, e.g. Nève who, otherwise, shows quite a rapport with the ancient Vedic inspiration and vision:

> The being of God . . . has in fact been destroyed by the naturalism of the Vedas, and of this one truth, the foundation of all others, there remains but a vague and imperfect notion, falsely applied by the intelligence of the nations. Divine existence is not denied, but as it is related to what is not itself, as it is dispersed in the multitude of things, so to speak, in losing its nature, in getting out of its laws, it ceases to be. The God of the

> Rishis and of the pastors has hidden himself in the forces and phenomena of the external world; he has multiplied himself as intelligent principle animating the various manifestations of life-giving light.[8]

The divine life cannot lose its nature even when fragmented in its manifestations. The God of the ṛṣis is both in the phenomenal world as in the human heart, as also beyond, in the highest empyrean, as the following verses show:

> Men find him in the hidden place (I.67.2)

where He manifests as the flame of aspiration, of inspiration, of exaltation, of 'highest immortality', the hidden place of the human heart where men fashion their *mantras*; but also as the 'Supreme Overseer'

> He who in the highest empyrean surveys [the world] (X.129.7)

both verses showing the immanence and the transcendence of the Vedic conception of Deity.

The inspiration of the ṛṣis does not stem from an anthropomorphic Deity that stands outside his creation. As mentioned in the introduction, the sages conceived Deity first in terms of an Absolute beyond human speculation, beyond the sway of the opposites that rule this world, as 'that' which 'breathes breath-less-ly by Itself', through whose creative contemplation (*tapas*) the flame divine of *kāma*, desire, is aroused,[9] whence the creative God or Demiurge manifests, as a result of which the universe is projected from within without as a series of unfoldment, and consciousness emerges from its matrix of unconsciousness, life out of the waters of space, light out of darkness; the gods being but the more or less personified energies and units of universal intelligence that pervade, mould, activate and dissolve the manifested world. No loftier conception of Deity, transcendent and immanent, can be found in the Bible, a Deity that is never dragged down to the level of human anger, frustration, revenge and hatred; all the gods forming the many faceted link between that which is beyond human conception, and the human inkling of some supernal power at the heart of the universe; each god representing one aspect of this power; each bearing a share in the construction and destruction of the ever moving panorama of the universe. If we fail to grasp these two important points, the Vedic vision of cosmic order will remain as insignificant to us as it has been to the Western scholars.

But the hymns of the Vedas testify to the ṛṣis' concern at viewing the pageant of the universe, its vast movement and what lies behind it as its 'noumenon', in one vast sweep of vision. The Absolute had to remain on its pinnacle of sacred silence, with only a few references to it here and there. But the one divine life-being-light manifesting through nature's varied forms, and in that sense deemed fragmented by the Belgian scholar quoted above, is nevertheless the link between Deity and man, that link embodied for the ṛṣis in Agni the flame seated in the human heart. True, the mirror of nature, however infinitely varied and beautiful, is but a distortion and a limitation of the infinite life that animates all things; yet the many images in which the bards clothed their apprehension of the divine life, whether as 'flame', 'sun' – 'ātman of what moves and moves not' – the life-giver and bestower of insight, a 'light', that 'celestial light lodged in the human heart without which nothing can be done that is done' (Yjv. 34.3), or as 'fine-winged bird' (*suparṇa*), or as 'untiring herdsman continually journeying through the world clothed in diverging and converging forces' (I.164.31) – these images do not militate against the grandeur of the vision and inspiration leading to revelation, except in the eyes of the prejudiced who cannot cast off their old skin of the anthropomorphic deity; nor do they detract from the vision of Deity – as grand in his immanence as he may be sublime in his lofty transcendence; but on the contrary they testify to the ṛṣis richness of expression and experience. Their vision of the beauty and all pervasiveness of the divine life as expressed through the gods that make the world alive and throbbing with a supernal quality which we in our conceptual thinking have lost, is in every way equal to that of the psalmists:

> Thy force and might and passion neither these birds in their travelling can attain, nor these waters ranging restlessly, nor they who hedge in the hugeness of the winds.[10]

All the powers of nature are here taken as witness that even they, however great, cannot compete with the might of Varuṇa. This takes the idea of God far beyond any natural expression, but nature's phenomena are used as a prop to the imagination to point to a transcendent power.

It is easy to explain the Vedic vision and inspiration in terms of natural phenomena, both being so closely interlinked, and to leave out of consideration, and even at times of translation,[11] such telling words as heart, mind or consciousness, intuitive insight (*manīṣā*), visionary insight (*dhītiḥ*), and to take such words as seers, inspired poets, sages, or others as sun, fire, lightning, waters, ocean, river, rock, cow, at their most superficial value and literal meaning. This is simply evading a fundamental issue, for the appeal of the poet is repeatedly to his insight in his heart where he finds the 'ageless flame' and the hidden light placed there as in all creatures. In their quality of inspired seers (*vipaścitaḥ*) they

> behold, with their heart, with their consciousness, the bird anointed with the magic of the asura (X.177.1).

Within that ocean – the ocean in the heart – they behold the source of his rays. Who is that bird anointed with the highest divine power? As J. Gonda points out:

> It is hardly questionable that in X.177.1 heart and ocean are identical. 'The inspired ones . . . see the bird (i.e. the light of insight and inspiration) with their heart; the inspired poets see (it) in the ocean.'[12]

The bird which J. Gonda explains as the light of insight and inspiration is the human spirit that is capable of soaring to the very heart of divine inspiration.

The 'bird', the 'ageless one', the 'unborn', the 'flame' divine, the solar splendour, are all expressions of a divine visitation or apprehension which the poet experiences in his heart, i.e. at the very core of his being.

> In the hidden place (of the heart) men find him when they have sung their mantras which they fashion in their heart (I.67.2).

Eagerly they seek out the 'seat' of his rays.

> Watching over the ageless one within their heart the wise seers take him in various ways to his abode (I.146.4ab).

Longingly looking into the 'ocean', they find the 'sun made manifest for men' (I.146.4cd).

Contemplation is in the heart and may have reached its acme during the rite of sacrificial offering, for many of the poets' visions resulting in praise-songs are offered to their chosen deity during the rite, and furthermore it is said:

> The wise seers watch over their inspired intuition refulgent as heavenly light, in the seat of *ṛta* (X.177.2),

ṛta here being the ritual as reflecting the cosmic order.

But the fountain source of inspiration welling through the heart and the object of contemplation were, for the ṛṣis, both the gods and the cosmic order as reflected in their consciousness; from the cosmic order the vision is said to issue and it finds its ritual counterpart in the sacrifice.

Vision, resulting in eloquence and prayer, is viewed as shining forth; as granting light or enlightenment; as an interior illumination whose mainspring and ground of being is the cosmic order. There is thus a visioning of aspects of the cosmos and a sense of rootedness in the latter that nothing can shake. These are then the three basic elements of the notion of Vedic revelation:

1 the *visionary quality* of the perception of the bard who combines insight into transcendental realms and verbal eloquence as his special gift, that which makes him a ṛṣi;
2 the *luminous character* of the vision; it is a glowing revelation of what is hidden to the senses; the revelation is uplifting, joy evoking and at times grandiose;

3 *its source is ultimately in the cosmic order*: the root of reality is the divine harmony mirrored in the human psyche, in the heart of man, perceived, understood and given expression to by means of the bards' songs, praises, incantations.

The many remarks as to the luminosity of the vision give the Ṛgveda its peculiar sunshine quality. Thus:

> From the seat of *ṛta* shines forth the visionary insight (X.111.2a).

Likewise from that same seat issues the prayer. That the latter also 'shines' is mentioned in an invocation to Agni:

> O tireless knower-of-generations! Bring us the *word* that fulfils, that it may shine to heaven (VI.16.36).

The word that fulfils opens out the intuition, lets in the light, like a flash of lightning. Prayer in the Ṛgveda, is *brahman*,[13] the evocative power of the *word*, *vāc*, the eternal sound or vibration which the ṛṣis discovered as dwelling in themselves as in the heart of the universe. Seeing and hearing are closely linked. A further verse tells us:

> By the song born of *ṛta* the sun shone forth (X.138.2d).

Interpreted psychologically this would mean that as a result of the poet's song which was inspired to him in his contemplation of the cosmic order, whether this be the harmony of nature and the heavens, or their ritual counterpart, light was made in his consciousness, hence the sun shone forth. The sun, representant of light and heaven, the organ of sight and insight generated by the gods themselves, the 'eye' through which they view the finite and the infinite and look upon the deeds and misdeeds of men, is the trophy won by Indra (X.43.5) and the goal set to every bard, a goal occasionally reached:

> Gazing upon the loftier light beyond the darkness, we have reached Sūrya, divine among divinities, supernal light (I.50.10).

One verse sums up the idea of prayer as the sacred word

that may lead to the beyond, as the basis of the praise-song that invigorates, as a result of which the 'sun' becomes apparent:

> Heed the brahman; be invigorated by the songs; make the sun manifest (VI.17.3).

The sun of truth (*satyaḥ sūrya*) or the 'true sun', the principle of enlightenment buried deep within each human being, will manifest as enlightenment as a result of the power of the holy *word*, prayer, of the exaltation of the bard as he sings his praises to the god.

Again:

> Forth glow the hidden visions of their own accord (VIII.6.8a),

causing the seers themselves to glow. They are compared to the glittering flames of Agni who is said to gleam peerless with bright visions (or revelations). With this divine vision (*devīm dhiyam*) the poet enwraps his god.[14] There is a pyschological hint here of deepest interest. The god is generally the cause of the vision, but the vision gives shape to the god, clothes him, brings him closer to the poet. Agni, himself 'immortal light among mortals' (VI.9.4) 'brings thought to birth within the heart as light'.[15] As already noted men find him in the 'cave' or 'hidden place' (guhā) 'whilst they give shape to their *mantras* in their heart'.[16] This 'hiddenness' of the vision and its luminosity transport it to the level of divine illumination. Agni is asked to 'release' the inspiration for the singer just as a cave opens out.[17] The inspiration may thus be regarded as deeply buried and in need of the flame divine to be able to burst forth on the horizon of the poet's consciousness. As J. Gonda points out:

> It is worth emphasizing that the belief in a light which, being suprahuman in origin and penetrating into the heart of the inspired poets, illumined their mind, was the complement of the conviction that these poets owe their praeternormal knowledge and their religious and political inspiration to 'visions', that they 'saw' the truth about

> the deeds and the power of the gods which they formulated in their hymns. These authors, indeed not rarely alluded to an internal light which is in the heart of the poet or to which he gains access in his heart. . . . This light is brought into connection with the sacral word of the inspired poet. The expression 'bearing light in the mouth' . . . is in 10.67.10 applied to the poets who magnify Bṛhaspati [the Lord of prayer].[18]

The ṛṣi indeed claims to discover a 'steady light', 'swifter than mind' stationed among all creatures for all to look on. And his ears strain, and his eyes speed after this light 'set within the heart' (VI.9.6). This light was described in the Yajurveda as the

> deathless flame in living beings without which nothing whatever can be done
> that immortal essence whereby all is comprehended, that which is and that which will be hereafter.[19]

Agni and Soma indeed are credited with the bestowal, upon the sage, of that ability to read into the beyond in order to find the essence of being. What significance Agni and Soma had for the ṛṣi will be investigated in the section on sacrifice. We have to realise that all nature was alive with divine life for the ancient poet, hence, e.g., his choice of 'fire' as divine messenger or soma as the draught of godly bliss. It is we who have weaned nature of its divine life and see nothing in it but mechanical phenomena.

Soma, the Lord of seer-poets, he who, wisest craftsman, knows the hidden nature of the 'cow's secret name', is said to give voice to

> The vision of *ṛta* and to the inspiration of prayer.[20]

> Lord of all those who see by Heaven's light, he roars forth the vision of *ṛta*.[21]

He grants that eloquence necessary to voice that vision which is received by those who are celestial-eyed, i.e. who can gaze into the beyond, the inner heaven, and word this insight as praise to the gods. Praise is the bard's prayer, as much as request or supplication.

Not only is the light quality of the vision emphasised, but also its state of being hidden; the latter may refer to the sudden irruption before the seer's inner gaze of something which till then was unnoticed, invisible, or to the fact that the vision remains invisible to others, to those not gifted with seership, as the bard is. Our

> forefathers discovered the hidden light and with effective mantras brought forth the dawn.[22]

What is this 'hidden light' (*gulhaṃ jyotiḥ*) which elsewhere is described as a 'steady light' fixed within the heart? The above verse epitomises the whole question of Vedic visionary insight and its expression; it contains the secret of the link between light and sound. By means of the power of sound (the mantras) the ṛṣis caused their consciousness to be illumined. This implies their well-known claim to 'see the Veda' as they chanted their invocations.

Not only is the inspiration shining like celestial light, but the seers keep watch over it, brood over it in the seat of the ritual. Possibly they meditate upon it until such time as they can express it to their heart's content. Meditating upon the flame divine in its three aspects

> there I saw the lord of being with his seven offsprings (I.164.1).

Remarks as to the 'hidden nature' or 'concealed knowledge' or 'mysterious and secret name' of cows or of the flame divine, to an 'inner light' swifter than mind without which nothing in the world can be, attest a hidden lore and an esoteric language current among the ṛṣis. So Agni, the discerner, reveals to the bard a 'wise insight' or 'intuitive perception' (*manīṣā*) as one who detects 'the hidden footprint of the cow' (IV.5.3). This Agni's 'loftiest secret name' (X.45.2) is known to the poet who refers to himself as a 'simple-minded' man to whom Agni the wise 'declares the hidden meaning'. Elsewhere (VII.87.4), to the bard who has the required 'wisdom' (*medhira*), Varuṇa declares the 'three times seven names of the cow'. This is recondite language indeed, but when we remember that Aditi, in whose depth all things are born, is called the 'cow' by 'simple-minded men', we may have a glimpse into the three times seven levels of manifestation inscribed in the ṛṣis's mind as the three times seven names of the cow, names being seats or levels of being.

The flame divine is man's lofty inspirer, amongst others, but it seems to be the closest to man, the immortal in mortals, dwelling in the human heart:

> What thing I truly am, I know not clearly; mysterious, fettered in my mind I wander. When the first-born of Holy Law approached me, then of this speech I first obtain a portion.[23]

The bard complains of the limitations of his mind, he is bemused as he tries to find out his true self, and indeed as comes out in certain hymns, the significance of this all, this vast panorama of conflicting and yet ultimately harmonious powers. But when the first born of the Cosmic Order, Agni, visits him in his contemplative moments, then he obtains a share of the 'word', the divine *vāc*, the inspiration made concrete through the *word*, the human expression. The theophany of the flame divine opens out the insight of the bard and loosens his tongue. Agni, whose origin can be traced back to the contemplation of the Absolute (X.129.4), the kindling flame of *tapas* who is thus first born of the Cosmic Order, is the 'immortal light midst mortals', the 'steady light' fixed within the heart, the 'wide ocean that holds all treasures and gazes out from our heart'.[24] Upon the bard Agni confers the 'hidden knowledge' (IV.5.3). Such communion reached its pinnacle when Agni,

> the lordly herdsman of the whole universe, the enlightened one, has entered into me, the simple.[25]

These are all expressions of upliftment of heart and thought, of inspiration, theophany, divine penetration of the human being and complete absorption. The cause of the theophany is certainly not of a low order, but out of it the bards would exclaim:

> We have drunk Soma; we have become immortal;
> we have gone to the light; we have found the gods
> (VIII.48.3).

Contemplation, method and tradition

But the method of inducing visualisation and of objectifying the vision so that it can be expressed, if it was formulated, is not explained.[26] Only hints are given out. That the practice of meditative absorption goes back to Vedic India is attested by some of the remarks found in the hymns. Scholars generally refuse to admit that 'yoga' was known to the ṛṣis of the Ṛgveda period. This is ignoring the essence of yoga – concentration, meditation, contemplation. The injunction to 'harness your mind' or 'contemplate aright' is a proof in point. That contemplation lies at the root of the ṛṣis' vision is evidenced by the many references to their vision. It is not because no mention is made of *asanas* and other terms peculiar to certain techniques of yoga, it is not because concentration seems to be made upon exterior deities, an outer projection instead of interiorisation of the thinking principle of consciousness, that there is no trace of yoga, or meditative absorption, in the Vedas. The very claim to seership implies ability to see 'within' which itself implies a contemplative mood. That the aims of yoga and of the ṛṣis may have been different is another question, the one purporting to lead to liberation from the shackles of *this* world's existence, the other to knowledge of this cosmos rather than freedom from its trammels. The means of achieving both converge on concentration, meditation, contemplation, absorption. One has to consider this question in terms of the particular values of the Ṛgvedic civilisation which were life asserting, as against the later trend as in the Upaniṣads which emphasised the quest for truth as basically an inner focus of one's whole being.

Superficially the seers may appear to us to be extroverts; they chose what to us are external symbols such as the elements – wind, storm-god, dawn, fire, sun – to worship.

Western scholars were at a loss when faced with gods to whom a 'physical basis' could hardly be ascribed or with gods that seem to fuse into each other. We have to bear in mind that what was external to prehistoric man was also internal, for ancient man did not differentiate as sharply as we do the external from the internal and tended to project outwardly what he found within himself; or it may be that we have lost our interior vision and so cannot understand that others should have had one. Nature was alive to ancient man with his own livingness; he had not lost contact with his inner being and saw far more than what the senses tell us. The Chāndogya Upaniṣad shows the link between the outer and the inner in a way which provides a perfect commentary for the ṛṣis' vision as the basis of their knowledge of the universe:

> What then is that which, dwelling within this little house, this lotus of the heart, is to be sought after, inquired about, and realized? As large as the universe outside, even so large is the universe within the lotus of the heart. Within it are heaven and earth, the sun, the moon, the lightning, and all the stars. What is in the macrocosm is in this microcosm (Chānd. Up. 8.12.3).

So the seers are said to

> harness (their) minds, harness (their) visionary-insights.[27]

The harnessing of the mind hints at a certain kind of mental training at least in so far as concentration and direction are concerned, and this is borne out in Ṛgv. III.38.1 where the poet admits to

> straining after an intuition, bounding forward like a spirited well-yoked horse.

So the god Varuṇa is asked to

> sharpen the resourcefulness and dexterity of him who is attempting (to express) this vision.[28]

There is here evidenced a definite attempt at an intellectual coming to grips with visionary insight. The mental qualities are sharpened in order to give shape to that which is seen inwardly at the level where words have no meaning, but must now be translated into language communicable to others.

Furthermore the vision is described as

> This vast, seven-headed thought-provoking-vision (*dhīḥ*) born of ṛta (the cosmic order).[29]

The vision has not only seven dimensions to it which are never described, but emerges from the cosmic order, the ground of all being, and is therefore inherently part of it; it is the human psyche's apprehension and reflection of what is in the cosmic order; this vision, or the fact of visioning, was discovered by the ṛṣi's ancestor and it became an heirloom for other ṛṣis who also lay claim to it. Two of the seven dimensions may pertain to the realms of light and sound, for the ṛṣis 'saw' and 'heard' the Veda, they tuned in to the infinite rhythm that underlies the movement of all things. The importance of this verse lies in the fact that it points to an ancestral vision to which the ṛṣi has access, to that vision as immersed in and emerging from the cosmic order, hence as an expression of it, or a reflection in his mind to which he gives new shape through words, by means of his hymn of praise or invocation.

> This new insight, worthy of attention and cherished of the many, I fashion in its own proper time.
>
> Or
>
> Perfecting it in-accordance-with-the norms of *ṛta*, I give form to vision.[30]

When the vision has been perceived in depth and brought to the foreground of consciousness it is then fashioned, 'meted out', expressed into words. But the vision is qualified both as 'ancestral' and as emerging in the poet's consciousness at dawn; it is born of old and yet comes to the poet each day.

> Born of the first daylight, watchful, recited at the sacred

> function, robed in auspicious raiments of light, this is
> our ancestral vision, born of old.[31]

The poet receives it as the archetypal vision granted to his ancestors from of old, newly grasped by himself and framed into song. Hence its value as revelation and as the secure ground of truth, hence its reliability. It is the divine heirloom, revealed anew each day and yet hereditary among the ṛṣis, that which makes them ṛṣis.

Steps in the process of contemplation are enumerated. A trusting approach to the god who is credited with certain powers that open up in the human being the required channels is evident. Thus Varuṇa is said to 'make the brahman', i.e. to cause the divine power locked in prayer to manifest:

> Varuṇa makes the brahman; to the path-knower we turn;
> in our heart he opens out thought; a new (manifestation
> of) *ṛta* must become known.[32]

Varuṇa knows the way; he is the one to whom one may turn in full confidence. Thanks to him thought opens out in the 'heart'. The juxtaposition of thought (*matis*) and heart (*hṛd*) implies depth of insight rather than superficial, or just intellectual appraisal. By such means a new understanding of the meaning of the cosmic order becomes possible.

One poet grants us a deeper glimpse into the manifestation of vision. By means of contemplative exertion (*tapas*), that flame-power that awakens spiritual insight, he admits to perceiving the condition wherein visions were aroused. This all important hint points to the fact that the capacity for entering into such a supra-sensuous state of supra-mental apprehension was furthered by *tapas*, a much misunderstood word implying a strict discipline of mind and body leading to a state of consciousness out of all relation to the normal daily state, a limitation in one sense for a further expansion in a deeper sense, of which 'heat' is only the physiological effect. It is not heat that arouses vision but the fervour of contemplation that itself manifests in the body as heat.

Certain graphic images are found in the Ṛgveda from which some conclusions may be drawn as to a method or technique of concentration being used by the ṛṣis. Thus a comparison is made between the seer's activity and that of a builder:

> as a skilful craftsman builds up a chariot
>
> as a carpenter have I been absorbed in meditation[33]

both implying a certain degree of concentration, directed thinking and a particular aim in view. In Ṛgv. I.88.4 the Gotama seers compare themselves to vultures that for days circle round their prey, which turns out to be that visionary knowledge which, like birds of prey, they relentlessly pursue. The factors of time and exertion are hinted at by the simple words 'for days'. Another hymn images a well whose lid has to be pushed open by means of the sacred word (brahman); its top is of stone and only the Lord of Prayer, Brahmaṇaspati, succeeds in doing so. It is filled to the very brim with honey.

The ṛṣi stresses that Indra and Varuṇa originally gave to the seers 'inspired insight' (*manīṣā*), the power to express thought (*vāco matim*) verbally, and revealed knowledge:

> O Indra, Varuṇa, when in the beginning, you gave to the seers insight, thought (capable of) verbal expression and revelation, I perceived by means of the fervour-of-contemplation, in what conditions the wise ones who performed the worship projected (the visions).[34]

The word used for 'fervour-of-contemplation' is again *tapas* which shows its fundamental meaning as intrinsically linked to meditation and visioning and as expressive of the technique of the ṛṣi. By such knowledge, reached through such meditation, men were raised to the status of ṛṣis.

> Praising the cosmic order and contemplating aright

the patriarchs, the sons of heaven, having attained the status of sages were able to

> devise the first forms (or norms) of worship.[35]

What the right kind of contemplation could be is not

explained but it is by such meditation that they were inspired to institute ways of paying homage. This is primarily the sacrificial ritual. Praising the cosmic order may thus mean celebrating the rite commemorating the original act of creation. The equivalence of ritual offering and cosmic order will be examined in the section on sacrifice. Here we can but note the importance ascribed to the act of meditation whereby the patriarchs became sages and instituted norms of worship.

These same ancestors are also stated to have

> found the light; (and) expressed it by means of *dhīḥ*[36]

Literally, they 'fashioned' the light by means of their visionary insights; in other words, they gave form to the illuminations they received; through songs, through rituals, they expressed what they visioned and instituted certain rituals. In mythological language, in 'breaking open' the 'rock' (*adri*), or, as we might put it, in breaking through the 'unconscious',

> they made for us a way to lofty heaven.[37]

Such verses hint at the patriarchs' achievements; they outlined the path whereby men might reach to where they themselves had gained access.

The visionary insight is the possession par excellence of the wise (*dhīra*); their wisdom is based on their vision; the *dhīḥ* opens up inroads into what otherwise would remain plunged in darkness; the wise man can penetrate into the very heart of life. Vision-inspired men are 'artisans' (*kāru*) of illumination, they not only strive after wisdom (*medhā*) but also give shape to it.

> These artisans, these vision-inspired-men have aspired after the gaining of wisdom.[38]

The Ṛgveda looks back to a golden age of seership and a living relationship between man and god, whose continuity is regarded as unbroken. The ṛṣis refer again and again to the fact of their seership. A well-marked trend in the hymns not to break with tradition is reflected in the hints as to the handing down from generation to generation of certain secret knowledge through which the bards traced their line of descent to an actual figure or to several patriarchs, to those ancient

> followers of the law, steadfast in the law, who furthered the course of law, the Fathers, Yama, moved by contemplative-exertion.[39]

Thus the ṛṣis call upon Agni 'as did the Bhṛgus, Manus, Angiras' in days of old. Another declares that 'our forefathers' guided by Soma

> received as their share, among the gods, a jewel.[40]

whereby they were made wise (*dhīrāḥ*). This is the same jewel as that 'heaven-allotted treasure' to which the patriarchs awaken when they behold the gods in every dwelling (IV.1.18) – the gift of visioning. Such continuity both at the physical (in e.g. the performance of specific rituals) and at the spiritual levels, in the receiving and handing down of insights, was a proof to the bards of the firm foundation of their science of seership:

> Not breaking the links, let us, seeking help follow after the powers of our forefathers.[41]

> We speak by-reason-of our descent (birth) from the ancient Sire.[42]

This Sire being Soma, the giver of enlightenment and eloquence, the great Inspirer, the ṛṣi here makes plain his heritage. The insight that brings the illumination made the poet a sun:

> Having received from my father the intuitive-perception of truth I was born like a sun.[43]

We may gather from the examples brought forward that it was through and because of their visionary powers, their 'seership', that the poets of the Ṛgveda arrived at their particular view of life, devised their particular forms of worship and framed their particular insights into songs, often through the colourful language of myths, images, symbols, but also in the simplest language as evidenced in the great creation hymns, and that because of the hereditary quality of their seership they felt secure in the 'truth' of their insights and 'revelation'.

From the available evidence we may conclude that there are stages in the reaching of the goal which is illumination or visionary insight into the beyond; that each step may have been marked out as a technique of concentration, meditation, vision, but is not mentioned in the Ṛgveda. It is however stated that in the 'fourth' degree of 'prayer' (brahman) the seer found the 'sun', the fourth degree implying three preceding ones, all of which are found explained in the Upaniṣads. The culmination of the visioning seems to have been absorption into a transcendental state of being, described in various ways, such as finding the sun, or being 'born a sun', or as 'god possession' (X.164.21). In such heightened state of illumination the ṛṣis fathomed out and developed their conception of the cosmic order. Their claim not only to knowledge of the laws of the universe, their understanding and application to daily social norms, but also to knowledge as to how best to bring the visions forward to the consciousness and express them as inspired utterances is thus founded on that 'revelation', which comes to man through 'visioning'. The latter is an opening up to, and a penetration into, the deepest structure of the macrocosm, and by repercussion, the microcosm. This insight into the cosmic order the inspired sage constantly desires. The vision in turn fosters the means of developing it through the application of the law. This brings us to the fundamental issue of this study, namely the examination of the many faceted *ṛta*, the vision of the cosmic order, which in course of time was transcribed in various ways with different emphasis, hence the myths, riddles and the philosophical speculations.

TWO

The meaning of the word *ṛta*

Definition of the word *ṛta* and sphere of *ṛta*

Certain Sanskrit words are not amenable to one specific meaning, but as in the case of *ṛta* are many faceted. The particular word, *ṛta*, has certain definite connotations such as reality, law, order, norm, basic structure of the universe, universal harmony, truth, depending on whether one examines the conception from the physical – including the luminaries in the sky and the biological regularity of terrestrial phenomena, the repetition of cycles, the balance of nature – or from the ethico-social, ritualistic or religious points of view.

Basically, *ṛta* concerns the *dynamics of manifestation*, the process of world unfoldment at all levels. In the Vedic vision the universe manifests in accordance with an inherent law which is the very basis of its structure; it unfolds not in a haphazard way, but in a strict order, a progression, all other laws being but the development of and, therefore, subordinate to this one fundamental law. The concept of *ṛta* as the expression of law in activity which we would call the law of becoming, or transformation, is contained in the very root of the word itself: √*ṛ*[1] which means to move, to go. Hence *ṛta* is the mode whereby the unmanifest becomes the manifest, the transcendent the actual, chaos cosmos. But, as grammatically it is a past participle, it also contains, as V. M. Apte points out, the notion of ' "something gone over correctly" or better "the settled or ordered course of going".'[2] Hence the order of things, both in their becoming and in their settled configuration, both dynamic and apparently static. These two aspects should be kept in mind in any assessment of the meaning of *ṛta*: the overall order results from the working out of all in accordance with the inherent law of being which itself is unchangeable.

It is apparent that this Order is not fully manifest in the phenomenal world and that creation is the process of actualising in this world that which exists at a transcendental level. That it exists at that level will be shown in our examination of the metaphysical origin of *ṛta* as conceived by the ṛṣis. In its transcendental sense *ṛta* may be taken as a kind of blueprint according to which all the world will be made manifest, this activity being itself the dynamic expression of *ṛta* as reflected at more and more differentiated levels of the cosmos of which the last, most differentiated and concrete level is our space-time world where man appears, and in his ignorance or ill-will impedes the working of the law which nevertheless finds its reflection in man's socio-ethical and ritualistic activity. Thus is *ṛta,* the one law, expressed, more or less successfully, at every level.

As explained by H. Lefever, the term *ṛta* contains three basic elements in its connotation: 'activity, order and law. . . . They are not always equally stressed.'[3] The 'activity' or 'dynamic' aspect of *ṛta* is often lost sight of, the static side, in the sense of unchanging order, being emphasised, resulting in a disregard of the vast movement of the cosmos that is the very expression of the law.[4]

L. Renou's definition takes account of both aspects:

> *Ṛta,* which for convenience sake can be translated by order (cosmic order and moral order) or by Law, is, more precisely, the result of correlations, the product of 'adaptation', of the 'fitting together' between the microcosm and the macrocosm.[5]

'Adaptation' and 'fitting together' are aspects of the law of becoming or transformation which operates in full at the secondary level of manifestation, that level where the *devas* come into action and 'measure out', 'prop asunder' and 'establish' heaven, earth and the mid- or inner region, all in accordance with the law inherent to their very being which is but a reflection of the law inherent to the very core of the universe. As the manifested world is a world of activity, as everything is on the move, as it were, we shall naturally examine *ṛta* in the light of this movement, *ṛta* in its dynamic aspect rather than static, the latter referring merely to the unvarying constancy of the movement, the regularity and predictability of all change.

The sphere of *ṛta* is all inclusive; there is no activity whether metaphysical, natural, or human that is not covered by it. Thus

> heaven, the wide expanse of earth, profound obeisance and more wondrous adoration (X.92.4ab)

fall under the domain of *ṛta*. In this verse are succinctly brought together the three most important aspects of *ṛta* and its field of expression:

First, *ṛta* as it manifests in heaven which, metaphysically, connotes the domain of the gods, their interaction and influence, hence the first differentiation of the transcendental *ṛta*. If 'heaven' is considered in its physical sense of sky, then this domain of *ṛta* pertains to the zodiac.

Second, *ṛta* as it manifests on earth in the orderliness of the natural processes, the time sequences, the balance of nature, the vast sweep of evolving life. Both Heaven and Earth are said to

> dwell in close union in the womb of *ṛta* (X.65.8b).

They are fields of manifestation for life through which the one all-inclusive law of becoming expresses itself. Both realms, being described as united through the one law are, by implication, shown to be working in harmony towards a common end.

Third, *ṛta* as it manifests in human life both in so far as its socio-ethical norms and its religious life are concerned, the latter expressed by the words 'obeisance' (*namas*) and 'piety' or 'adoration' (*aramatiḥ*). This implies that homage and obeisance are brought together in the sacrificial ritual where man re-enacts the universal process of *give* and *take*, and adds his own contribution. In obeisance and worship he aligns himself to the universal order, pays his tribute to the one Law and becomes the living agent of the *ṛta*. In this way he makes himself whole, integrated, true. He is true, hence *ṛta* as truth. This is religion, this is the ritualistic aspect of *ṛta*.

To sum up:

> 1 *Ṛta* is at the *natural level* the one law that underlies the basic structure of the universe in accordance with which all evolves, all suffers transformation, all gives and takes and from which derive all other laws; hence, at this level *ṛta* is the law of becoming, but that becoming which conduces to harmony.
>
> The panorama of the world shows a conflict and a tension between polar opposites, but also a movement in togetherness and a co-operation. All are striving towards the light, towards unfolding their potential, towards fuller life. The many forms that enshrine life, in their constant appearing and disappearing on the mundane scene, may give the impression of disorder, of opposing forces that strive against one another to gain the upper hand. The trees in the forests vie with each other for the light. So do men in the jungles of cities, for success which is their light. The strongest survive or rise above the others. The weakest disappear. The survival of the fittest is a law of nature.
>
> Underlying this fever of activity is the law of becoming, of transformation by means of which all unfold, adjust and evolve. Becoming, transforming, and in the process adjusting, are the very expression of evolution, the dynamics of life. The vast movements of the heavens and of nature's varied forms are outer expressions of an underlying orderly process that sweeps all forward unrelentingly, whose regulator is the Law, *ṛta*. The very constancy of the universal movement along regular lines

marks an underlying stability, which is the settled order, *ṛta*. Motion and stability meet in the *ṛta*.

Thus is cosmos constantly wrought out of chaos, the lower order endeavouring to mirror the higher, the 'pattern in the heavens', in spite of all oppositions and imperfections. This is the law.

2 At the *socio-ethical level* where humanity is concerned, *ṛta* finds expression as *truth;* that one truth which in the context of human norms can be translated as integrity-integration; man fulfills himself in as much as he lives truly and can therefore integrate himself in the cosmic order.

3 At the *religio-sacrificial level* man himself reflects the one sacrifice with which the cosmic order is equated, a constant give and take of all the units of the universe, an eternal exchange which is sharing, which itself is rooted in the law of transformation, of balance, of harmony.

To this order all beings contribute in greater or less measure, whatever friction occurring from non-co-operation or opposition being finally offset by the greater pull towards harmony. Nothing can really override the one law. Through the agency of gods *ṛta* as the 'pattern in the heavens' is made manifest at the natural level and through the co-operation of men in the norms and ethics of human society. Man's sacrificial rite is itself seen as his microscopic attempt at re-enacting the macroscopic drama of the dynamics of *ṛta*.

In thus analysing the word *ṛta* into its component aspects, we should not lose sight of its 'unity', its oneness, its whole. One order rules the cosmos at all levels, the same order manifesting differently at different levels. As H. Lefever rightly stresses

> *Ṛta* in nature, as in the sphere of sacrifice, refers not to any particular 'order', e.g. the order of the sun, but to the natural order 'in general'. In nature, as in the sphere of sacrifice, there is only one *ṛta*, which is manifested in the different natural 'orders'. . . . In its application to human conduct, no less than to sacrifice or to natural phenomena, *ṛta* is a unity. . . . Different obligations may

apply to different men and circumstances but it is the same moral order – *ṛta* – in every case. So those who cleave to *ṛta* are held righteous by all mankind (V.67.4).

Not only is *ṛta* a unity in each of the three spheres mentioned, it is a unity throughout them all. Each of the three 'orders' – sacrificial, natural and moral – is a manifestation of the same universal *ṛta*.[6]

The physical image of *ṛta*: the zodiac

Although nineteenth century scholars tried to explain almost every Vedic conception as based on an observable natural phenomenon, or on a particular sensory object, there was little attempt, with regard to *ṛta,* to emphasise the physical basis of the conception to the exclusion of other aspects, beyond the recognition of *ṛta* as meaning the regularity of the movement of the celestial bodies, of time sequences, the succession of the seasons, and so on. This was itself an evaluation of the word at its most material level, i.e. as it manifests in the world of physical phenomena. It was thus summarised by Max Müller:

> this Rita . . . meant originally the firmly established movement of the world, of the sun, of morning and evening, of day and night . . . the spring of that movement was localised in the far East . . . its manifestation was perceived in the path of the heavenly bodies . . . that right path on which the gods brought light out of darkness became afterwards the path to be followed by man, partly in his sacrifices, partly in his general moral conduct.[7]

The meaning of *ṛta* as moral order captured the attention of scholars and so they emphasised that aspect at the expense of the others. With a few exceptions among the poets and the philosophers, nineteenth century Western man could not soar much beyond the compass of specific moral values by which criteria societies were judged and usually found wanting, little knowing that this very morality on which he prided himself so highly, which he prized as the supreme arbiter of civilisations, was going to be completely toppled down within just a few decades. Hence the vision of the cosmic harmony beyond human moral evaluation, such as we get from the scattered hints in the Vedas, was beyond nineteenth century compass and ken. In the twentieth century we have at least realised that morality is only a human value judgment and therefore only a small part of whatever picture we may form for ourselves of the vast cosmos. The Vedic vision of the cosmos is beyond human moral values although it includes these in its breadth as we shall see in the section on the moral aspects of *ṛta*.

An interesting development of the nineteenth century tendency to interpret Vedic conceptions as always rooted in a physical or visible phenomenon is found in the twentieth century scholar V. M. Apte who, in his article 'The Zodiac – the physical basis of *Ṛta* (Cosmic Order)'[8] brings forward arguments for the validity of such an interpretation. His thesis is that in addition to the number of passages where *ṛta* undoubtedly means law, order, right,

> there is an equally large number of passages there which betray (what we may call) the primary or the fundamental physical sense of the word which represents . . . the natural or physical basis of the very abstract . . . conception of *ṛta* as 'Cosmic- or World-Order.'[9]

Ṛta, according to Apte, thus refers to

1 the path of the luminaries of the zodiac
2 the world order
3 right
4 rite.

Since to the Vedic mind all things were interconnected and indeed interdependent, the physical world would thus have its counterpart in the non-physical and vice-versa. *Ṛta* – if the idea of the zodiac could have been part of its several meanings – could then have been considered the visible reflection of the invisible spiritual expression of the universe, but the point should not be stretched too far.

Certain words such as 'womb' (*yoni*), 'bottom' (*budhna*), 'cavity' (*khā*), 'summit' (*sānu*), 'wheel' (*cakra*) are used governing *ṛta* and expressive of the heavens in a physical sense; the word 'seat' (*sadana* and *sadas* and *dhāman*) and others also often seem to refer to a particular location of *ṛta*; all can also be used figuratively. The 'path of *ṛta*' may be taken in both a figurative and a literal sense – the path of righteousness as well as the path in the heavens along which the dawns, the sun and the other luminaries travel (and the course of the sacrificial ritual). Some verses show this ambiguity and may be taken in two or three senses:

> For the broad (sun) a wider course was made manifest,
> the path of *ṛta* has been extended by means of rays
> (I.136.2ab).

Slightly less obvious is the following:

> The path of *ṛta* is for crossing safely to the other shore,
> the channel of heaven has become visible (I.46.11).

The channel of heaven which becomes visible is the path of *ṛta* which can be used to cross to the beyond, the word *ṛta* here possibly bearing three connotations, the track of the heavens along which move the stars, that invisible way which leads to the other shore, as well as the sacrificial ritual which re-enacts the cosmic drama. The paths of the planets in the heavens are expressive of that eternal recurrence, that peculiar rhythm, which is one aspect of the cosmic order and which the bards tried to capture and express in their rituals.

This 'path of *ṛta*' Dawn 'follows correctly',[10] not missing the 'directions' or cardinal points. This evidently refers to the path that dawn follows in the heavens and that this is the path of *ṛta* is quite significant for it is the path laid out by the law and also laid out spatially in the heavens. Nor does she 'trespass on the *dhāman* of *ṛta*'.[11] V. M. Apte takes this verse in a literal sense as meaning that Dawn 'never misses the habitation (*dhāman*) of *ṛta*',[12] i.e. she illuminates the zodiacal belt. It is, however, open to both literal and figurative interpretations. It can equally mean that Dawn never infringes the law, rule, or statute in so far as it applies to her nature, eternally recurrent illumination which is part of the vast cosmic rhythm regulating all things in manifestation. Each day she appears, that is her appointed task.

The sun described as 'the pure and beauteous face of *ṛta*' (VI.51.1c)

> besprinkles with honey the womb of *ṛta* (X.68.4a)

an action which points to the golden luminosity extending over the cavity of the sky at dawn, so that the sun, as it appears, could be thought of as that radiant face that illuminates the canopy of heaven which itself is the womb of *ṛta*. Sūrya, 'glowing in the seat of *ṛta*', 'shines on the summit of *ṛta*'.[13] For him, have the 'Adityas cut open a course'.[14] The Sanskrit word for 'course' is *adhvan*, used in the accusative plural, meaning either road, course or orbit, but it is also given out as 'zodiac' in Monier Williams's dictionary. The orbit of the planets is laid out by the gods, so their pathway in the heavens is both the law and the zodiacal belt.

The most convincing example brought forward by V. M. Apte to show that one meaning of *ṛta* is the zodiacal circle is found in the great riddle hymn of book I by Dīrghatamas (I.164.11):

> Around the heaven revolves the ever unaging twelve-spoked wheel of *ṛta*.

This could and has been interpreted as the sun circling round the sky but in this case the twelve spokes would be difficult to explain. The verse lends itself to the idea of the heavens conceived as a wheel with its twelve spokes or divisions circling round forever through the twelve months of the year; 'there' (*atra*), i.e. within that wheel, the verse goes on,

> Seven hundred and twenty sons abide in pairs.

The word 'pairs' gives the clue to the meaning of seven hundred and twenty sons; we may presume that these refer to the three hundred and sixty days and three hundred and sixty nights.

Furthermore in verse 2 of the same hymn, the 'one-wheeled chariot', 'three-naved',[15] 'ageless' and 'unobstructable', 'whom the seven yoke', is said to be 'drawn by the one steed with seven names'; 'there', i.e. within this wheel, 'all the worlds take their stand'. This cannot refer to the sun only as most scholars interpret 'wheel', since all the worlds are not located in the sun, but rather to the zodiacal belt which contains them all or possibly only to our solar system; the sun itself being the steed and the seven other steeds, occasionally mentioned, the seven planets often referred to in antiquity.[16]

In the same hymn, verse 48, this 'single wheel' – the emphasis on *cakram eka* is quite significant – is described as having 'twelve fellies' and being 'three-naved' as in verse 2:

> Fixed together therein are three hundred and sixty spokes that (turn) undeviatingly.

The eternal recurrence of day and night, of the seasons, the years, assures the irrevocable march of time and the ordered process of nature which nothing can disturb. This 'selfsame wheel' is drawn by the 'dappled' or 'brilliant' courser, the sun itself, *etaṣa*,

> intent on revolving it hither, yoked to the poles (VII.63.2cd).

This verse, as indeed verse 2 previously quoted, is usually interpreted as referring to the sun being drawn by a horse and the wheel is taken to be the sun itself. But the interpretation of the sun being itself the courser that draws the wheel or circle of the heaven is equally valid. Sūrya is at other times represented as the leader of the coursers, the latter possibly referring to the planets.

All these verses seem to bear references to *ṛta* the cosmic order as mirrored in natural phenomena both in the heavens and on earth, the zodiac being but the image of the orderly motion of the universe, hence the image of *ṛta*. But this is only one of the many facets of *ṛta*.

THREE

The metaphysical origin of *ṛta*

Introduction

We may speak of the Ṛgvedic vision of creation, using the word 'creation' in a general sense, but we should bear in mind that in the context of the Vedic understanding of beginnings the idea of *creatio ex nihilo* as envisaged in Christianity does not apply. The ṛṣis thought of the origin of the universe in terms of a projection into manifestation of that which lies latent within the One, the word used being *visṛṣṭi* or *visarjana* from √*sṛj* to let go, to emit. It is rather like an unfoldment from within without in which process we notice three definite stages which imply three levels or world orders:

1 The primordial or transcendental level where

> that one which flame-power kindled to existence emerged (X.129.3)

and *ṛta* manifested, born of *tapas* (X.190.1). This is the transcendental *ṛta*, the blueprint of the cosmic order. Ultimate oneness is a characteristic of the Ṛgvedic vision to which Western scholars paid little attention:

> One whole governs the moving and the stable, that which walks and flies, this variegated creation (III.54.8cd).

> That which is one has developed into the all (VIII.58.2d).

From this One differentiated the two poles of manifestation, poles which we would term positive-negative, spiritual-material, but which the Vedic seers first expressed as *asat* and *sat*, non-being and being, or the unmanifest and the manifest, or the potential and the actual, as Aditi and Dakṣa born from each other, and then as Dyaus-Pṛthivī, Heaven and Earth, and Apas-Tvaṣṭṛ, the Waters or feminine potency and the creative power of nature.

2 The level intermediate between the spiritual and the physical, where the gods appear on the stage and through their interaction establish, give shape to and rule the manifested universe, or lower order, in accordance with the law implicit in their nature as at the heart of the universe, *ṛta*.

3 The concrete level, or space-time phenomenal world, where man rules and upsets the divine equilibrium. Here the tension between the opposite forces manifests as conflict, hence disharmony. It is the constant task of gods and men to restore the harmony and thus actualise at the earthly level that which is real at the ideal level.

Each of these orders will be examined in so far as it represents a field for the manifestation of the law of change, transformation, order, equilibrium – *ṛta*. It will become apparent that, although the law is one and the same at all levels, it finds a different expression according as the levels of life become more and more differentiated.

We have first the all-pervading law of transformation, of becoming or differentiation which could be simply called *evolution*, that which comes into play when *sat* emerges from *asat*, when the *be* of the universe is no longer a potentiality but an actuality. This law then manifests upon the different rungs of the ladder of the cosmos that could be summed up thus:

At the *primordial* level the law of becoming is seen in the differentiation of the One into the two poles of creation, which two in turn differentiate into three, the world of Vedic cosmogony, which three become in due course the playground of the many; in this interaction of two complementary poles, in this play of opposites, we glimpse the root of all the opposites such as light and darkness, positive and negative, good and evil and similar dichotomies without which a universe cannot be.

At the *secondary* level the law of transformation manifests through the action of the devas in so far as they are the law's agents in the further differentiation of the universe, in their propping asunder the fields of Heaven and Earth, their separating the various realms of nature and laying down the norms of human society and ritual, acting in accordance with the law inherent to their being. The universe is the expression of the devas' harmonious activity. Here we find the field par excellence of the dynamics of *ṛta* where the law of harmony is seen in its fullness, the very expression of the pattern of Order.

At the *third* level the effect of the law of becoming becomes more complex; for opposition to its harmonious working is set up by human agency through ignorance, wrong choice or sheer ill-will. This creates an aspect of the vast Order which is peculiar only to its relation to humanity: the law of righteousness as reflected in the socio-ethical-ritualistic side of the one universal law and its ramifications at all levels and man's relation to it. We cannot regard this last aspect as the all important side of *ṛta*. It is but one of the facets of the many faceted law and one that manifests only in relation to humanity. Truth, righteousness, justice, are human value judgments which find their counterpart in the universe as harmony. These are the human reflection of the Cosmic Order; the fact that they are not prevalent in human society but nevertheless ideally envisaged and aspired to by the sages proves human capacity to 'vision' and human failure to 'mirror' or establish the law securely.

What is constant as the characteristic of *ṛta* in the three fields of manifestation is the law of transformation, ordering all things towards one harmonious whole, a cosmos, hence *ṛta is essentially the law of harmony*.

It will be also apparent that, if we are to divide the cosmos into neat compartments, for the sake of analysis, a certain amount of overlapping is inevitable. Interdependence and interlinkedness are a marked feature of the universe which did not escape the perception of the ṛṣis. Neat, separate compartments do not exist in nature. Thus for example the question of the poles of manifestation examined in connection with the first differentiation will have to be reconsidered in connection with the second level where the pairs of deities in their interaction further separate or delimit the world. Similarly, the question of the gods, whose activity is considered at the secondary level, will have to be taken up again in connection with humanity because of the close interaction, both at the ethical and ritualistic level, between both classes and their influence on each other. Some of the examples chosen from the Ṛgveda may also have to be taken up again in various sections to be examined from their various aspects as one and the same verse may bear several meanings.

The word *tapas* often used in the creation hymns as well as with reference to the spiritual exertion or self-kindling by *tapas* of gods (Indra) and men (the ṛṣis) is one of those important key-words upon the right interpretation of which depends our whole understanding of the process of creation. It is usually translated according to its literal, outer meaning, as 'heat' (from √*tap* to burn) or as 'austerity, penance' when referring to men, with little consideration for its figurative, metaphysical meaning which is of primary importance as the latter covers not only the process of creation at the highest level that man has ever been capable of envisaging, but also the ṛṣis' creative contemplation and realisation of the whole cosmic unfoldment. Enquiry into the genesis of the whole taxed their visionary powers which they developed by means of *tapas*, that intensive concentration upon an inner focus which arouses the creative fire in both gods and men.

Ontologically, *tapas* represents that cosmic energy which keeps the universe moving from the beginning to the end of the cycle. The transition from the Absolute, the One beyond being or *sat* and non-being or *asat*, merely hinted at in the first and last stanza of the *nāsadīya* hymn to the manifested One described as 'the overseer' (*adhyakṣa*) who, even as the Supreme One in the highest empyrean, may not know the ultimate secret of Reality, occurs through the inherent creative drive of *tapas*. Through *tapas* the Unmanifest becomes the manifest. It is the primordial creative act par excellence.

Psychologically, *tapas* refers to a specific kind of inner exertion, an intense spiritual focusing which takes place in deep meditation, a contraction to an innermost point of contemplation, and a subsequent expansion whereby the creative flame is aroused at the highest level of awareness, the physiological effect of which only is heat. To emphasise the effect at the expense of the cause evidences complete ignorance of the whole process and indeed failure to grasp the very essence of contemplation. The poets in these creation hymns are using the analogy of yogic meditation to describe the divine creative act. This contraction-expansion arouses the supreme energy locked within the One and personified in the Ṛgveda as Agni. Not content with mere *tapas*, the bard of the last but one hymn of book X of the Ṛgveda qualifies the word *tapas* with the participial adjective of the verb to 'inflame', so that we have the superb picture of the divine contemplation blazing to incandescent manifestation:

> Universal order and truth were born of blazing *tapas* (X.190.1).

Through the power of *tapas*, through the power of *kāma*, of that primeval desire for existence, for manifesting, for 'being' to emerge out of latency or 'non-being', that desire which provides the bond between the unmanifest and the manifest, was the universe 'projected' out of potentiality into existence. There is but One causeless Cause, the *tad ekam*, that One. This is all too often lost sight of, hidden as it is behind the multiplicity of the gods and their activities. Very few hymns refer to this because nothing more can be said about it, our world being one of multiplicity, and the understanding of nature's infinite variety being really the concern of the ṛṣis, that which taxed their vision most.

The primordial differentiation

To form a picture of the ṛṣis' conception of the origin of the universe in its metaphysical aspect, as against the myths that were developed to explain the same idea, we have to consider several of the creation hymns. At the primordial level of manifestation, we find posited 'that' which appears to be the Vedic conception of the Absolute. In the celebrated *nāsadīya* hymn (Rgv. X.129) this Absolute or 'that' appears as beyond human speculation, beyond attributes, although the human mind cannot help but qualify it in some way or other, such as that

> whose shadow is death whose shadow is immortality,[1]

or as 'undifferentiated depth', or 'ocean profound, unfathomable', that which breathes or pulsates by Itself, though without breath – paradoxical indeed, but ultimate being is beyond all mental logic and all known experiences.

> The Unmanifest was not then, or the Manifest; spatial depths or heaven beyond were not. What encompassed, where, who nurtured it, What ocean, profound, unfathomable, pervaded?
>
> Death was not then or immortality. Neither night's nor day's confine existed. Undisturbed, self-moved, pulsated the One alone. And beyond that, other than that, was naught.
>
> Darkness there was; at first hidden in darkness this all was undifferentiated depth. Enwrapped in voidness, that which flame-power (*tapas*) kindled to existence emerged.
>
> Desire, primordial seed of mind, in the beginning, arose in That. Seers, searching in their heart's wisdom, discovered it as the bond between the created and the uncreate.

> Their vision's rays stretched afar. There was indeed a below, there was indeed an above. Seed-bearers there were, mighty powers there were. Energy below, will above.
>
> Who knows the truth, who can here proclaim whence this birth, whence this projection? The gods appeared later by this world's creation. Who then knows how it all came into being?
>
> Whence this creation originated; whether He caused it to be or not, He who in the highest empyrean surveys it, He alone knows, or else, even He knows not![2]

Within 'that' arose what could be described as the 'creative power', the upward shooting force (*uttānapad*), as a result of *tapas* and *kāma*. So the ṛṣis traced the origin of manifestation to the emanation or projection (*visṛṣṭi*) from the One through that flame-power aroused as a result of contemplative exertion (*tapas*) or through desire (*kāma*), of a propelling force which as Agni is said to pervade all things and contain all powers.[3]

> Both tapas and action were within the mighty ocean. Tapas arose from action; that did they worship as highest.[4]

Here in this Atharvaveda hymn, both *tapas* and *karman* (action) are seen in the 'ocean', i.e. in the depths of space. We might consider this verse as referring to creative contemplation as hidden in the great sea of cosmic life, i.e. in the unmanifest. The action is the arousal of the creative energy which itself leads to further action. The action of both *tapas* and *karman* is reciprocal.

The conception of the ultimate reality as an impersonal 'that' on which there can be no speculation, is now stepped down so to speak to a supreme, masculine overseer in the highest empyrean, who himself may not know the ultimate secret of creation, for beyond him stands the Unmanifest, the Absolute in whom alone the secret may be locked. In the *nāsadīya* hymn that overseer remains nameless. In other creation hymns Varuṇa stands as supreme and sovereign ruler over all, master or overseer of the ṛta, the 'Omniscient' who is said to

> behold all things transcendent, past and future[5]

who is the 'hidden depth', 'in whom rests all wisdom'. Or, the All-Creator, Viśvakarman, 'our Father', the 'Lord of vision', is another presentation of this 'overseer', who is earlier than Heaven and Earth and the gods, who

> all-seeing, all-facing, all-embracing, all-pervading, with his arms, with his wings altogether whirled forth, he the One God, generated Heaven and Earth (X.81.3).

The whirling forth of the Creator, the One God (*deva ekaḥ*) brings to mind an original vortex of energy circling round and round, aroused to activity through the *tapas* of the Supreme, and bringing all things forth. Fire was at the beginning and fire pervades the whole of creation:

> Universal order and truth were born of blazing tapas (X.190.1).

The origin of *ṛta* in its metaphysical sense is hinted at in the above verse, part of a short but revealing three stanza hymn of creation. This brings out a point of far reaching importance in the Vedic vision, namely that truth (*satya*) and order (*ṛta*) are inherent in the very structure and process of manifestation. Creation occurs in accordance with the inherent law of being. That is truth (*satya*). Hence the Vedic vision of the world is one of universal order, that transcendental order manifesting at the very beginning of this world's projection, being the blueprint according to which all orders of manifestation will be modelled. From this all-kindling tapas the right order of things, the process of transformation, *ṛta* is set in motion, in other words, the course of existence started to roll onwards. So Abinash Chandra Bose, in his *Hymns from the Vedas*, wrote:

> '*Ṛta* and *satya* as ends, and *tapas* as the means, form the foundation of Vedic culture and religion.'[6]

The law came into action with the very first throb of manifestation. Thence was born time (*saṃvatsara* – the year) which rules all animate creatures with its orderly sequence of day and night, time the 'ordainer' (*dhātā*) that shapes or 'images forth' the stars and the heaven and the earth and the intermediate region. The orderly unfoldment of the universe is, in Ṛgv. X.190, stated in a few masterly strokes:

> From the billowy ocean was born time ordaining days and nights, ruler of all appearances.
>
> As before, the Creator fashioned forth the sun and moon, and heaven and earth and the mid-region and also the empyrean (X.190.2.3).

To the minds of the seers there was a difference between *ṛta* and *satya*; both are named separately as they emerge from blazing *tapas*. Order and truth are two aspects of the one harmony. *Satya* (from *sat*) refers to that which is, the inherent being or being-ness of anything; *ṛta* (from √ṛ to move, rise, go) refers to the 'course of things'; being in manifestation, in activity, the dynamic aspect of being, its working out, not only the visible order but what underlies it, the universal structure which is cosmic harmony. One may infer that *satya* is the passive or static expression, and *ṛta* the active or dynamic expression of Reality. J. M. Koller takes one step further along this line of interpretation when he declares;

> There is no difference between the being (*sat*) of reality and its function (*ṛta*). . . . This identity of being and function (*sat* and *ṛta*) is one of the profoundest aspects of the Vedic vision . . . *satya* here (X.190.1) refers to the inner being or truth of reality, and *ṛta* to the functioning of that reality.[7]

But the equation can only be conceived as existing at the transcendental level. As H. Lefever points out: '*Ṛta* includes *satya*, but its content is wider. Hence what is "not *satya*" is also against *ṛta*, and so *anṛta* is regularly used as the opposite of *satya*.'[8] *Anṛta* is untruth, falsehood, that which goes against the order and is therefore false. The main difference between both words could be summed up thus: 'subjective truth' as against 'objective order', inherent being as against manifest law, *veritas in essendo,* essential truth as *satya* and *veritas in rei,* contingent truth as *ṛta.*

Satya in its absolute meaning is described in the Ṛgveda as

> luminous, lofty, established on the unshakeable
> foundation of heaven (X.170.2ab)

and also as 'the base that bears the earth' (X.85.1a). Essential being, or beingness, as the basis of the world, finds its manifested expression as order in *ṛta* whose root is thus ultimate being. So the ṛṣi states:

> Firmly fixed are the foundations of ṛta
> shining in beauty, manifold are its beauteous forms
> (IV.23.9ab).

These forms are the manifestations of the world order, e.g. the perfect harmony existing between the essence of being, *sat,* and its activity or outer expression, *ṛta*: a thing will grow only in accordance with what is in its seed; so long as we ignite fire in the proper way that will produce the spark, it never fails to be kindled, for we have touched its inherent nature. This is the law, this is *ṛta.* Therefore fire is kindled in accordance with *ṛta,* the tides ebb and flow in accordance with *ṛta,* the sun rises, extends its light and sets in accordance with *ṛta;* the stars move along their ordered pathways, the seasons come in regular succession, the dawns, the days and nights, birth, growth and death all follow in ordered sequence and form part of the rhythmic pattern of life, the eternal rhythm of the universe, ordered activity, *ṛta.*

> By *ṛta* (the law) have the herds (of the universe) entered
> the *ṛta* (cosmic order) (IV.23.9d).

All the worlds move in accordance with the law, and all things in those worlds also proceed in accordance with the law:

> In accordance with the law is eternal nurture outpoured
> (IV.23.9c).

> Dawn the maiden infringes not upon the domain of Order
> and day after day proceeds to the appointed-place
> (I.123.9cd).

No god infringes the common statutes. All natural activities are effects of an inner harmony which is the touchstone of the Supreme Order.

These examples show both the static and dynamic aspects of *ṛta*; what is ordained and therefore fixed as the law, is immutable; nought can override it; that which brings about what is ordained, the dynamic activity expressed through those agents of the law, the devas working in accordance with the law which is basic to their very nature, that is constantly in movement; hence the continuous actualisation of that law, the eternal transformation of all things in accordance with a specific pattern being inherently the settled order.

This orderly process that works according to the inherent law of being can be reduced to a fundamental rhythm: in-breathing and out-breathing, withdrawing and out-going, involution and evolution, action and reaction, contraction and expansion, diastolic and systolic, give and take, or in the quaint language of the hymns

> weave forth, weave back (X.130.1d).

Hence the institution of rites to commemorate and translate into human terms the universal order; hence the bold imagery and equations in which the ṛṣis seem to have delighted:

> To law (belong) the deep and wide (Heaven and) Earth,
> Supreme milch-kine, to law these outpour their milk
> (IV.23.10cd).

Dualism inherent in manifestation: the question of evil

The question of opposites, of contrast, of friction, or in human ethical terms of evil, considered in general, is contained both in the philosophical implications of *asat* and *sat*,[9] which two polarities stand at the origin of all, and in the dualism that underlies manifestation, the opposition and therefore friction that is inherent in any existence.

The primeval differentiation that heralds the unfoldment of the universe is seen in the *sat* or being, the 'be' of manifestation emerging from the *asat*, that which is not and yet contains the 'be' potentially, emerging or aroused as a result of the flame-power of *tapas* (X.129.3).

> From the *asat* (non-being) the *sat* (being) was born (X.72.3b).

Metaphysically *sat*, from which *satya* 'truth' is derived, is the root of manifestation and its opposite *asat* is the not yet manifest; *sat* can also be envisaged as the actual expression of the potential (*asat*); it is the field of manifestation, of organisation; *asat* the field of the indeterminate yet potential, of chaos that will give birth to cosmos. *Sat* and *asat* in the philosophical hymns refer to two poles of the one state of being which itself is beyond all conception and is the substratum of all that is. The *nāsadīya* hymn points to a primeval state when neither the potential nor the actual, neither chaos nor cosmos, *asat* nor *sat*, was. There the poet attempts to express a state of perfect oneness, beyond time, beyond space, beyond the sway of the opposites, yet containing the all. At one remove from such heights of perfect oneness over which the human mind can hardly range, are the *asat* and the *sat*. And it seems that the former engendered the latter or caused the latter to differentiate itself, in order to realise itself in the concrete. When merged these two poles are shown latent in

> the highest heaven, in Dakṣa's birthplace, in Aditi's lap (X.5.7),

the latter gods, Dakṣa and Aditi, being not personifications of *asat* and *sat*, but simply poles of manifestation representing certain conceptions which the ṛṣis placed at the origin of all, the infinite or space represented by Aditi, and intelligence or the creative power at work through space represented by

Dakṣa; movement in the infinitude. The dualities *asat* and *sat*, the potential and the actualised that form the abstract foundation or 'woof' of the cosmos, the seeds of which are locked in the divine intelligence, in the infinite, are transformed at a slightly more personified level in Aditi and Dakṣa who themselves are reflected in those two poles of manifestation which all cosmogonies have personified in Heaven and Earth, Father and Mother, our immediate parents. Originally interlocked or fused until such time as they separate, Dakṣa and Aditi come out of latency when the *sat* emerges from the *asat*, as though this primeval differentiation were made manifest through them. They are thus considered the first or 'primordial parents' lying far back of Heaven and Earth who are the more immediate, slightly more objectified parents.

Creation, as we have seen, meant for the ṛṣis the projection into limited or conditioned existence of that which contains all *in potentia*. This implies a reflection and at the same time a constriction or distortion of the unlimited or infinite source. In this mirror-like distortion the unconditioned finds itself imprisoned by the conditioned, such conditioning being essential to manifestation for without form or framework there can be no expression. The action of the creative gods, at the secondary level of manifestation is, as we shall see, to separate (prop asunder, delimit, etc.) in order to define, hence organise all in harmony with the one law.[10] Every projection must, by definition, be inferior to its source as it is a conditioned expression of it and ultimately may involve its very opposite, namely the negation, through limitation, of absolute being: the plenitude of manifestation can only be measured by the extremes of its contrasts, hence light and darkness, good and evil, life and death, growth and decay. Absolute being is

> that whose shadow is death, whose shadow is
> immortality (X.121.2)

hence beyond manifestation, beyond the poles of existence, the sway of the opposites. Whence metaphysically *sat* and *asat*, *ṛta* as manifested order and *anṛta* as disorder, are the two sides of one Reality, of which the transcendent *ṛta* alone can be the pattern. The essence of being is unitary, that which is projected so to speak out of it will become a reflection and in its outermost projection its very opposite, real only in so far as it borrows its reality from its source, hence in a certain sense illusory. Opposition is an inherent law of the universe. So is attraction. There can be no manifestation without opposition, hence friction, separation, evil.

The forces of opposition, whether they be described as non-being-being, (*asat-sat*), chaos-cosmos, true-false, were not in the Vedic view as sharply defined and opposed as some scholars claim.[11] The Vedic ṛṣis were far too aware of the interconnection between all forces to place them into neat opposite camps. The interplay of opposing forces may involve an intermingling of these as is exemplified by the ambivalence in the personality of some Vedic gods, such as Varuṇa and Rudra who both deter and soothe, destroy and build up or heal. The ambivalence of these is a notorious factor to be taken into account in any assessment of the gods. On this subject L. Renou declares:

> These cosmic powers, precursors of the *śaktis*, do not constitute a system of clear-cut oppositions. In classical times Śiva, the terrible destroyer, could also be a kindly protector; similarly in the Vedic system, vast spheres of activity are controlled by ambivalent powers. A normally well-disposed divinity may take on a *ghorā tānu*, an awful aspect; Varuṇa is alarmingly liable to assume the aspect of Vṛtra. . . . Sometimes the ambivalence is an integral feature of the divinity, as in the case of Rudra.[12]

The differentiating power at work throughout the universe may be said to be personified in some of the gods, notably in Indra who 'repeatedly (*muhur*) makes the *asat* into the *sat*',[13] or the potential into the actual, so that from the differentiation of the *asat* and the *sat* originally fused, chaos becomes cosmos; or this same power assumes the grand figure of Puruṣa, the divine prototype of man who, born in the beginning, was sacrificed that from his remains the universe might be; but only one-fourth of him pervades the world, three-quarters of him remain invisible in the inner levels.[14] The word 'repeatedly' which qualifies Indra's action, points to the process of transformation that is constantly going on in the world. The emergence of the *sat* from the *asat* is evidence of the starting point of the law of becoming, the ever changing process of universal life which we call evolution, from the indeterminate to the determinate, from the unconscious to the conscious.

The two opposing forces, the constrictive and the expansive, are somewhat personified in the Ṛgvedic dragon, Vṛtra, and the Vedic demiurge, the former, elemental, primeval, undivided or 'without joints', undifferentiated chaos; 'sunk in deepest sleep', Vṛtra represents the static rock-like power of inertia and darkness as opposed to the dynamic force of Indra with his thunderbolt, Indra, dynamic intelligence, the light-bringer whose action is necessary to cleave through the status-quo in order that an ordered cosmos may be brought about; Vṛtra has to be mythologically 'slain', dismembered or rent asunder, i.e. conquered so that what chaos contains *in potentia*,[15] may be released, developed, evolved, expressed, that inertia may be surmounted, that the orderly universe, made up of more or less clear-cut, separate parts (or joints) or divisions, may come into existence.[16] This is creation, manifestation. The idea is similar to the immolation of *puruṣa*,[17] only, in the one case it centres on the side of inert substance, on the power of constriction which is forcibly rent that its potential may become actual; in the other, on the divine, constructive power, or archetypal, all pervading deity, willingly dividing or immolating its homogeneity. The mountain which contains all things also refers to this primeval chaos; it has to be opened up so that the waters of life it holds back can flow freely, or the cows or rays of light can be freed.

The original sacrifice or primeval violence set the pace for every subsequent unfoldment or differentiation so that manifestation in the Vedic view appears as dynamic interaction between opposing forces involving both destruction and construction, involution and evolution, light and darkness, bringing to birth and rending asunder, activity and inertia, expansion and contraction, good and evil, all being equal and opposite and no definite line of demarcation can be drawn between them.[18] The alternance between one polarity and the other, their interaction and oft-time opposition, the succession of unmanifest and manifest, of birth and death, growth and decay, waxing and waning, the constant transformation to which all things are subject, implying the destruction of one form for the construction of another, point to a law of becoming, of change and adjustment, to which all creation submits, a law which is one aspect of *ṛta*, the ordered course of things; all being subservient to that one law of transformation, *pariṇama*, the equalisation of the opposing forces being part of the ordered course of the universal pattern, hence *ṛta* the law of harmony, the cosmic order.

The differentiating process means association and dissociation, hence friction, attraction and repulsion, hence adaptation or adjustment, all exemplifying the law of change and of balance, for balance is inherent in the adjustment of the vital tension. The form in nature constantly adapts itself to the need. This adaptation-balance is one of the most important facets of *ṛta* and manifests at all levels. In this play of opposites or complementary forces, there may be resistance, hence friction, hence pain. Hence the origin of pain, or as some would call it, evil, is rooted in the very process of becoming. Hence the interaction of gods and demons found in all cosmogonies. Hence the notion of cosmic evil which, to the present writer's mind, is basically the interplay and friction of the opposite polarities.

The continuous process of creation

Creation in the Vedic account is a continuous process. R. N. Dandekar notes:

> Creation is not a single definite act – it is regarded as ever proceeding . . . creation usually implies transformation of the potential into the actual or the establishment of cosmos in place of chaos. It is indeed in this latter sense that the *Ṛta* is often characterised as the source of the phenomena.[19]

'One action today, one action tomorrow' (VI.24.5a) rather depicts the idea of action succeeding upon action, like the dawns that never fail in their appointed task. After the *sat* emerged from the *asat* and some of the great gods had become active,

> Thence did divisions follow into existence; thence this (all) from the creative power (X.72.3).

The emergence of *sat* from *asat*, as we have seen, marks the process of unfoldment which also means 'demarcation', 'separation', 'moulding', 'measuring', the action of measuring in order to establish being typical of the gods. Chaos is becoming cosmos, the world order has come into being, *ṛta* is manifesting as a world organising power. To the ṛṣis' mind the world unfolded in accordance with a certain rhythm, expressing certain stages of manifestation, that rhythm being a facet of the orderly process of the law of harmonious becoming, *ṛta*.

The ṛṣis' vision seems to have already embraced vast cycles of action for in two hymns of the Ṛgveda and three verses elsewhere and also in the Atharvaveda hints are given as to successive emanations in time, or to use H. W. Wallis's phrase 'progressive stages in the growth of the world'.[20] For example the

> plants came into being three ages before the gods (X.97.1ab).

The Atharvaveda refers to the 'earth which was previous to this one' (XI.8.7). Heaven and Earth seem to be fashioned anew, or to be given a new lease of life.

The question of cycles or *yuga* is still moot, scholars refusing to admit that the ancient seers ever envisaged the vast cycles of time as are laid down in the Purāṇas. But the hints scattered in the Vedas seem to point to the opposite. If there was a planet earth before our own, if the plants came into being three ages before the gods, if there were different generations of gods, if Aditi presented her offsprings to the former generation of the gods, if the gods themselves *acquired* their immortality, or were granted it by Savitṛ, by Agni, or by Soma, implying that they were originally mortal, but are now immortal, *now* implying a different cycle, all these statements refer to facts or events that could only have occurred in previous cycles. For *now*, at the time of the seers, the previous earth is no more, and to know of its existence is considered to be 'wise in ancient lore',[21] the plants have been well established for ages, and the gods are the immortals *par excellence*. We cannot dismiss the whole problem by categorically stating that 'the early Ṛgvedic poets had no clear and definite notion on the matter'.[22] They were probably not inclined to expound the idea in detail, but their very idea of progressive emanations would automatically make them think in terms of cycles of generation and dissolution. The very words 'as before' (*yathā pūrvam*) in the short hymn of creation imply a cycle of time:

> As before [or as of old], the Creator imaged forth the sun and moon, the heavenly, earthly and intermediate realms, and also the empyrean (X.190.3).

Other questions related to the concept of cycles spring to mind, two of which are voiced by W. N. Brown to which he gives no answer, as any possible answer can only be in terms of 'ages' succeeding each other. Thus Viśvakarman the 'All-maker' is said in Ṛgv. X.82 to have 'celebrated or instructed the sages (*ṛṣis*) how to celebrate the first sacrifice, which was that of creation (and where, by the way, did these *ṛṣis* come from – there is no answer in the text)'.[23] Similarly with the Puruṣa hymn (X.90) wherein is recounted the myth of the primeval sacrifice of Puruṣa that from his remains the world might be, the gods performing the sacrifice with Puruṣa as oblation. The gods, as W. N. Brown points out, 'seem to be external to Puruṣa, but where they came from is not indicated in the hymn'.[24] The Puruṣa's original sacrifice refers to the beginning of this present world's dispensation when the fashioning of our world is thought of in terms of a sacrificial offering, the particular gods performing the sacrifice, as the ṛṣis of X.82 and the *sādhyas* of X.90, belonging to a former cycle, and therefore coming into being, closely clasped in the billowy deeps (X.72.6), in that primeval age of the gods when the *sat* differentiated from the *asat*. Thence was the sacrifice performed, itself a reflection of the divine contemplation of the One of the *nāsadīya* hymn when there was nought but the One.

The hierarchy of the gods is itself an indication of time sequence in their appearance and development. A verse from the Atharvaveda shows that at least some of the gods emerged to being when the world was not yet completely manifested, that they belong to the first age when the *sat* was not yet:

> Mighty in name are those devas who emerged from the *asat*; that *asat* is the one limb of the fulcrum . . . (Athv. X.7.25).

The actualisation of the *sat*, or of manifestation, is the work of the gods for

> from the unmanifest (*asat*) emerged the manifest (*sat*) in that primeval age of the gods (Ṛgv. X.72.3).

That the gods do not all come into existence at one and the same time, but that there is a sequence in their appearance, seems evident from certain Ṛgvedic verses:

> In the gods' mansion stood the foremost, and from their separation came the later (X.27.23ab).

In another hymn, the poet, intent on paying homage to all the categories of gods, reveals the hierarchical structure of the Vedic pantheon, the mighty and the older gods, the younger and the smaller:

> Homage to the mighty, homage to the smaller (gods) homage to the younger, homage to the elder (gods) (I.27.13ab).

Part II

THE GODLY LEVEL: THE DYNAMICS OF *ṚTA*

FOUR

The Vedic pantheon

Introduction: the interlinkedness of the gods

At the secondary level of manifestation, or that level which stands between the noumenal or pure abstraction and the purely phenomenal or realm of sensuous experience, the devas come into being and action.

Two hymns of creation make it clear that the gods are part of manifestation, they shape it, they rule it according to its inherent law which is also the law of their being, but they are not its original cause.

> Earlier than the heaven, earlier than this earth,
> earlier than the gods and the asuras, That indeed is.
> What is that seed primeval which the billowy-deeps
> conceived wherein all the gods appeared together?
> (X.82.5).

> The gods appeared later by this world's projection
> (X.129.6).

Vast and indeed at first glance somewhat confusing is the panorama that stretches out before our wondering gaze, but order can be found in this seemingly bewildering array of many gods whose shifting and overlapping roles tend to blur the picture. The poets in their cryptic references confront us with constant permutations which are but steppings down the various levels of manifestation that may lead the unwary astray and have puzzled scholars and created no end of misunderstanding. This permutation implies a process of objectifying powers at work in the universe at more and more concrete levels, all in accordance with the law of transformation. Hence this process whereby the abstract becomes in due course the concrete can be viewed as an aspect of the mighty dynamics of *ṛta*.

At this stage we find the divine, mysterious Power that resides at the core of the universe as in all its creatures, that is the ground of being rooted in the One (*tad ekam*), appearing fragmented in a multiplicity of shining or radiating, energising principles, the *devas*, the bright gods or 'shining ones' of the Vedic pantheon. Just as

> The One fair-winged Being, inspired poets, with their words, shape in varied ways (X.114.5ab)

so the One Power-Intelligence at this lower level appears to have split Itself up into the many intelligent units which we could consider as the *noumena* working behind the phenomenal expressions of nature, indeed of the whole universe, but which are also far more than that; for they are the regents of the manifested world as well as powers expressing themselves through men. Neither wholly personal nor totally impersonal,[1] and yet in some subtle way unique, these cosmic intelligences have individual characteristics and functions that mark them out each from the other, but their role and indeed whatever personality they evidence, or has been grafted on to them by the poets, do at times merge or overlap, so that they may appear as separate entities and yet not quite separate. Thus Brahmaṇaspati, Agni, Rudra in certain instances are identified or merge functionally into each other; thus Savitṛ, Sūrya, Pūṣan, Agni, are all expressions of one and the same solar power:

> I honour as the face of lofty Agni in heaven, the bright and holy light of Sūrya.[2]

Agni, the guest of men, the great mediator between gods and men, fire in the sun, fire in all things and fire in the heart of man, the same power that dries up and dessicates, that softens and melts, that builds up and destroys, Agni is all the gods because his dynamic, creative-destructive flame is within them all:

> In thee O son of strength, all gods are centred (V.3.1).

But he is also the whole of nature since he is found in the plants, in the wood, in the stones, in the waters, in the heavens (II.1). When the poet sings of Soma the giver of strength, the finder of light (VIII.48.15), he exalts both the ecstasy of life immortal, the divine experience of the god-entranced mystic, as well as the intoxicating power of that desire for existence that brings all things into manifestation, from the loftiest gods to the humblest blade of grass.

The interlinkedness of all the gods, manifesting as their solidarity, their essential righteousness, their concerted activity, is their peculiar feature and one that eminently marks them as the agents of the law of harmony, *ṛta*, by which, through which and in which they live and perform their varied functions:

> One is the mighty godhood of the shining ones[3]

is the refrain of a hymn that celebrates the devas and Agni, the 'universal king'. The emphasis on 'one' (*ekam*) is of peculiar importance in a pantheon of many gods; it points to the underlying *oneness* of these expressions of the Divine, the One mirrored through that oneness in its many facets or agents, the idea of which was quite lost in the later development of the Hindu pantheon and has been little studied by Weştern exegesis. Similarly, we are told in a creation hymn that

> when unfurled the mighty waters that generated the universal matrix . . . thence was that one life-breath of the gods breathed forth (X.121.7).

The underlying oneness of the gods is expressed as the *asur ekaḥ*, one life-breath that obviously differentiates in the many forms and functions of the gods, all nevertheless geared to one purposeful achievement

> all gods, one-minded, one-intentioned, unerringly

proceed to the one purposeful accomplishment (VI.9.5cd).

Each god can be worshipped as representative of the Divine, as a facet or an embodiment of it, which at any particular moment may be invested with the highest cosmic attributes or may include all attributes of other gods. The elevation to supremacy of one god or another at any time was called by Max Müller Henotheism. This was possible because the gods were regarded as expressions of the One, *deva ekaḥ* or *tad ekam*. Each god is a representant of the All-God: *viśvadeva;* so Savitṛ is All-God (V.82.7). So again, and here the example is from the Yajurveda:

> Agni is That, Aditya is That, Vāyu is That, Candramas is That (Yjv. 32.1).

Because of this Indra can be identified with Varuṇa (IV.42.3), Rudra with Agni (IV.3.1), Agni with all the gods (II.1); all the devas exist in Indra (III.54.17). The Sāmaveda refers to the 'Lord of heaven who is one' (372 *patim divo ya eka*). The Ṛgveda states:

> That which is one has developed into the all (VIII.58.2d).

This half verse clearly shows the ṛṣis' awareness of the underlying oneness not only of gods but of all things; all are expressions, individualised facets of the One, nameless Essence.

> One whole governs the moving and the stable, that which walks and flies, this variegated creation (III.54.8).

Classification

Various commentators have classified the devas variously. Thus Yaska in his Nirukta (7.5) followed the ancient Indian school of etymologists and placed the gods into the terrestrial, the intermediate and the celestial realms. Such grouping is based upon the clue given in the Ṛgveda itself.[4] Certain modern scholars, such as A. B. Keith, A. A. Macdonell and

P. S. Deshmukh, among others, followed in the wake of the Indian tradition, trying to fit the gods into three main categories according to the threefold division of the universe, the 'celestial' over which Sūrya presides, the 'aerial' or 'atmospheric' whose jurisdiction falls to Vāyu, or Indra whose growing prominence threw the older god Vāyu into the shadow, and the 'terrestrial' over which Agni is the regent. This naturalist division is confirmed in the Śatapatha Brāhmaṇa[5] but is unsatisfactory because of the overlapping in function and being of many of the gods.

So the ritualists inclined to view the gods according to their specific functions in the ritual. In this respect G. Dumézil,[6] among the modern scholars, worked out a theory that fits the gods into the three main human divisions of labour – the religious function (*brahma*), the ruling class (*kṣatra*) and the productive aspect (*viś*), of which, according to his estimation, the leading gods are Mitra-Varuṇa, Indra and the Aśvins respectively. Such divisions, arbitrary and artificial though they may appear when applied to such a fluid order as that of the Ṛgvedic pantheon, may help us to envisage the realm and functions of the devas in some kind of pattern, always keeping in mind their fluid interaction.

The point of importance is to examine the permutation of the gods and its significance as exemplifying the law of transformation in the wider context of the dynamics of *ṛta*. Their functions are distinct yet in some respect seem to merge into one another. Similarly, no god infringes upon the domain of another yet the demarcation of some domains remains fluid. In the gods' realm we catch glimpses of the human psyche's reaction to the wonders of nature; we are caught back into the fountain-source of our roots, our own psyche, there where our bond with Nature is at its strongest – for through the millenia psyche has grown nurtured by those divine principles which are themselves the fashioners of Nature, which psyche projected and anthropomorphised into the figures of the gods who in the Vedas appear as the bright gods of the Vedic firmament, 'sons of the Infinite', 'herdsmen of the Supreme-Law, whose decrees are truth', who ruled the cosmos in their perfect innocence, goodness

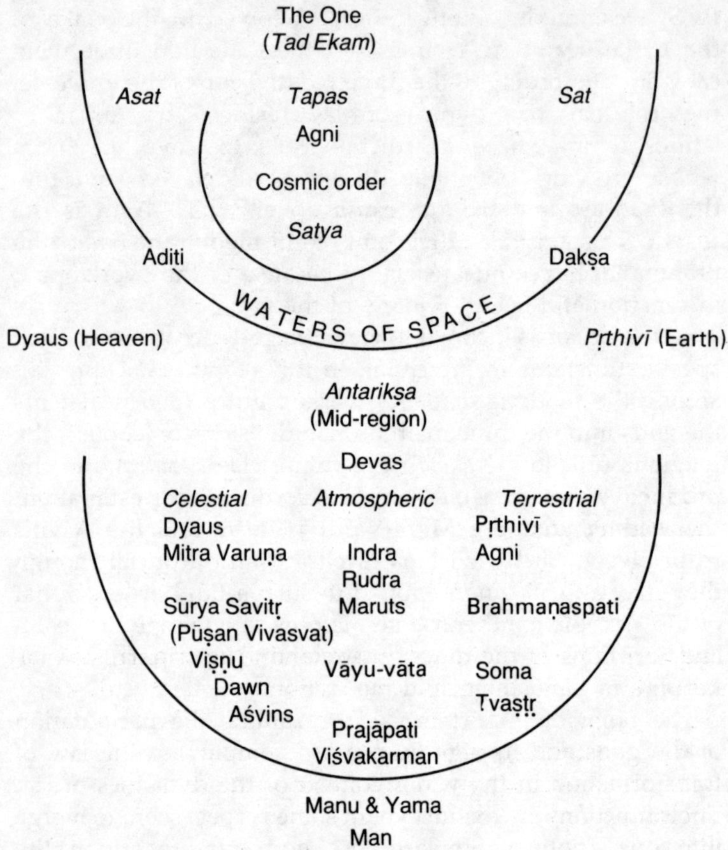

Figure 1

and purity, long before the human mind degraded them to its own petty level.

Birth of the gods

The birth of the gods is sung in one of the great hymns of creation

> The birth of the gods! Let us now joyfully proclaim it that in the ages yet to come it may be visioned in these exalting utterances (X.72.1).

The gods are born in the depths of space, of infinitude, *Aditi*, the 'all-bounteous', in the waters of space:

> In yon billowy-deeps ye gods stood closely-clasped; then from ye, as though from dancers, hot dust was whirled away (X.72.6).

The expression 'closely-clasped' is another instance of the underlying oneness of the gods, they emerged together, although as some other verses have pointed out, there are sequences in the gods' coming into being. But this particular category of gods seems to come into existence as luminaries whirling in a cosmic dance as they take shape in the depths of the heavens,which takes us right out into the inroads of the starry galaxies. The devas emerge to existence 'born of the waters'; the 'depths' or 'ocean' are referred to as the 'waters' in another verse where the thought of the ṛṣi is further expanded:

> Ye who were born of the waters and of aditi and of the earth (X.63.2).

The waters are the primordial substance, the original state of matter by which all things are shaped in the depths of space to manifest finally through the more concrete element, earth; in and through these are the devas fashioned.

The 'universal matrix' we are told was generated by the 'mighty waters'. They conceived the seed primeval wherein all the gods appeared. The waters are called 'mothers' in that they nourish like loving mothers; they produce Agni the flame divine (X.91.6); the lightning is their son; Savitṛ their offspring. In the waters is 'immortality', is 'healing balm' (I.23.19). The watery abyss, the great Mother, the Chaos of the Greeks, the Great Deep of Genesis, is not only the primeval *protomateria* out of which all the universes are fashioned, but also the womb wherein is hidden the elixir of immortality. In what man considers darkness, or chaos, the waters of space, lies concealed the principle of light, of order, of cosmos, of life-giving energy, the sun of enlightenment which it is man's task (personified by Indra) to release.

Another word for the celestial waters *ambhas* refers, when

in the plural, to gods, men, *manes* or *pitṛs*, and *asuras*. Varuṇa is the lord of the waters (*ambhaspati*); in their midst he moves, looking down upon the truth and falsehood of men (VII.49.3). This 'golden egg' (*hiranyagarbha*) of space contains all things.

Dakṣa – Aditi

We have seen how in the One are contained the two *asat* and *sat* and how these are latent in the highest heaven, 'in Dakṣa's birthplace, in Aditi's lap' (X.5.7), these latter two exemplifying powers at the same time impersonal and personal and only sufficiently manifested to be personified as Father and Mother. The Vedic seers did not draw any hard and fast line between the personal and impersonal aspects of their gods. In some hymns Dakṣa is personified, in others the word is simply used as an adjective.[7] So Agni and Indra are described as lords of *dakṣa* (*dakṣapatih* I.95.6), i.e. of that power of insight, and ability which is represented by the god. This might be a pointer to the antiquity of Dakṣa as a deity whose personification is very fluctuating in the Ṛgveda. In several instances he is numbered among the Adityas, although in Ṛgv. X.72.4–5 he seems to be prior to them. In the Śatapatha Brāhmaṇa (2.4.4.2) he is identified with the creator Prajāpati, thus keeping to his original role of primordial cause.

Dakṣa's name connotes the effective power that can realise in the space-time world that which has been thought out or ideated. Whatever degradation Dakṣa suffered in the later mythology, there is no doubt that in the Vedas he is a high God in his own right, though rather remote from mankind.[8] He figures in the cosmogonic hymns as that divine primeval impulse, that creative energy which, at the beginning of the aeons, brought about the first step in manifestation by differentiating from Aditi, the infinitude of space. He is that Intelligence that expresses itself through all nature's activities and that manifests through all her creatures who are born with a certain degree of it, great or small, the gods seemingly having the larger share. His is the 'mighty power' of which the most wise Mitra and Varuṇa are the sons. Hence, in accordance with the Vedic practice of personifying certain powers, Dakṣa is called their father, the gods being qualified as *Dakṣapitara*, to whom Dakṣa is father, or as the 'luminous sons of Dakṣa'. Mitra and Varuṇa who are 'the most wise sons of mighty Dakṣa', who rule with the might of Dakṣa, who are most 'dexterous',[9] best exemplify that constant fluctuation between the notion of a creative, almost personal power[10] in the universe and an impersonal attribute inherent in the creatures of the universe. Addressing them, the poet says:

> you have separated the chaos from the cosmos . . .
> through the zeal proper to your insight.[11]

Here is evidenced intelligent discrimination, that insight which sifts, orders, arranges, creates; that discrimination which conduces to right action, characteristic of the gods, and in varying degrees of men and beasts. This is personified in Dakṣa.

At the higher level this power could be envisaged as divine ideation of which the cosmic harmony is the blueprint; at the lower level as intelligence in practice, that which brings about in space-time dimensions what was conceived at the ideal level: by the creative power the ideal is slowly made manifest. This original connotation of Dakṣa, such as may be surmised from the way the word is used both as an epithet and as a deity, shows it far removed from the later god of Purāṇic cosmology.

Aditi[12] connotes the infinite depths of space, the matrix, womb or 'lap',[13] the generative feminine power, the primeval mother wherein all things come into being through the impact of that power called *dakṣa*, as a result of the act of *tapas*, that inner kindling that gave the original impetus to creation. She is the mother who brings forth the gods and within whom is all that has been born and all that will be born; the repository of the past, present and future, of the manifest and the unmanifest. This is well brought out in a Ṛgvedic verse:

> Aditi is Heaven, Aditi is the midregion, Aditi is mother

> and father and child, Aditi is all the gods, Aditi is the five races, what was born and what shall be born.[14]

She is universal space wherein all move and have their being, the waters of space, primeval matter, mother substance. To her goes the prayer of the Yajurveda:

> Aditi, mighty mother of just rulers and queen of those who follow Eternal Order, great ruler with a far extending sway, untouched by time, protectress, gracious guide, to you we cry (Yjv. XXI.5, Panikkar's trans.).

As in the case of Dakṣa, so in that of Aditi; the word also exhibits a fluctuation between a complete personification of freedom as a goddess and the simple attribute of freedom, a boon for which many a poet prays, as we shall see in the third part of this study.

Her glorious appearance in men's eyes is the Dawn (Ṛgv. I.113.19). She is 'not to be limited', 'far-pervading', 'supporter of all creatures',[15] 'celestial' (*svarvatim*), 'luminous' (*jvotismatīm*). Her three definite attributes are those of 'motherhood', 'spaciousness' and 'luminosity'. The basis of manifestation is that luminous space in which all things are born. Her qualities of light and space which may both be taken as symbolic of freedom are mirrored in Dyaus (from √*div* to shine) and Pṛthivī (from √*prath* to extend), Heaven and Earth respectively. The wide expanse of earth usually rendered by *urvī* is a direct reflection of Aditi; Heaven's light a reflection of her luminosity. Here we see how the concept of Aditi is stepped down and made slightly more concrete or tangible and objective in Heaven and Earth. She seems to stand behind those two as the great enfolding Mother, the celestial virgin, the still unmanifest bordering on the manifest and mirrored in Heaven and Earth.

The ṛṣis' vision of the universe as infinite is reflected not only in their conception of Aditi as unlimited expanse and freedom, but also in such expressions as the 'fathomless sphere' (*abudhne*) in which Varuṇa keeps erect the cosmic axis.[16] So the poet sings:

> In gain of wealth, we celebrate with praises her, Aditi by

> name, the mighty mother on whom this universe of life hath settled (Yjv. IX.5b).

From Aditi's body, the cosmic matrix, eight sons were born, so we are told in a Ṛgvedic hymn of creation; this could refer to our planetary system:

> Eight were the sons of Aditi born from herself.
> With seven she approached the gods;
> Mārtāṇḍa she cast away.
> With seven sons Aditi went forth towards the former generation.
> Mārtāṇḍa she again brought forth for the sake of generation and dissolution (X.72.8–9).

These are the original Adityas of the Ṛgveda subsequently multiplied to twelve, the great lords of light and omniscience. In the above stanzas, Mārtāṇḍa as the rejected son, is cast away for the sake of existence in time. The question remains: are the other sons not existing in space and time, hence under the sway of birth and death? But the Sun, as Sūrya, is also considered an Aditya and is depicted as drawn by seven horses or one horse with seven heads. What the connection between Sūrya and Mārtāṇḍa was in the thought of the ṛṣis is left obscure in the Ṛgveda.

Aditi's mother quality wins her the name of 'cow'. She complains:

> Weak-minded men have as a cow adopted me who came hither from the gods, a goddess; who, skilled in eloquence, her voice uplifteth, who standeth near at hand with all devotions.[17]

This verse hints at the connection between Aditi, space, and the *word*, the imperishable, sacred *akṣara*, that mighty eternal word born in the home of the 'Cow':

> When the first of dawns shone forth, was born in the home of the Cow, the mighty eternal word (III.55.1).

In the mythological language of the Vedas, the cosmic buffalo, or 'Cow'

> lowed as she fashioned the billowy waves (I.164.41).

She is the 'thousand-syllabled in the sublimest heaven' from whom stream forth the waters of life:

> The whole universe exists through the undying syllable that flows from her.[18]

> Two later scriptures afford an explanation and a revelation:

> In the beginning was *ākāṣa* which is the word.[19]

> The word is the syllable, first born of Order, mother of the Vedas, navel of immortality.[20]

Ākāṣa, the subtle essence out of which all things are moulded, according to later Indian philosophy, whose attribute is sound, is Aditi of the Vedas. Sound or vibration, as personified in the goddess Vāc, and ether or subtle essence of space as Aditi, become linked and merge in the one goddess Sarasvatī, mother of the Vedas, consort of Brahmā, the creator.

The process of permutation

The process of permutation which is part of the law of transformation and such a vital aspect of the dynamics of *ṛta* in the Vedic vision, is expressed not only as the stepping down of certain gods from the abstract to a more concrete level, but also both as the offspring begetting or consuming its parent and as the reciprocal birth of two principles. This has caused more puzzle than is warranted.[21]

'Being the son, thou becamest the father of the gods' is said of Agni in Ṛgv. I.69.1. Agni is generated by the gods through their mental abilities (*cittibhiḥ*), by the power of *dakṣa*, by the power of the word, but Agni makes the gods manifest, makes the gods immortal.[22]

> This is the law (*ṛta*); unto ye I declare it, O Heaven and Earth, the offspring on being born consumes his generators (X.79.4ab).

Here is admitted that *ṛta* is the great law of becoming, of change, of evolution. The parents, inherent in the embryo, disappear transformed in their offspring as the latter grows. So in the process of generation, evolution, in the vast course of *ṛta*, the new constantly destroys the old, taking from it what is necessary. This is the inexorable onward march of events, *ṛta* in its dynamic aspect, a Ṛgvedic conception which in later times was to be personified in Śiva the destroyer, the regenerator, the Lord of the cosmic dance.

The same idea is illustrated in the following verse:

> That which generated these things you will not find;
> something else has emerged to being from amongst you (X.82.7ab).

The generator has disappeared into those very things to which it gave birth. Yet in destroying the old the new once again begets the old though in a new form, for it can only beget that which is within itself and that is none other than what was given it by the old; this is the meaning of the offspring begetting its parent and this is the Vedic vision of eternal recurrence.

The Ṛgveda gives us two particular examples of the reciprocal birth of two principles. Thus:

> From the Infinite (*aditi*) Intelligence (*dakṣa*) was born,
> again from Intelligence the Infinite (X.72.4d).

Aditi at one level represents the infinite in its transcendental aspect, the infinitude of possibilities through and in which plays creative ideation, *dakṣa*, the interaction of which in its turn gives rise, at another level, to the more readily conceivable Mother-space notion that contains and nourishes everything, hence through which and in which all are born, the infinite source of life.

The famous Puruṣa hymn of creation wherein the divine or heavenly Man, or spiritual essence, is shown as being sacrificed that from his remains the many might come into being, provides us with the second example:

> From him (Puruṣa) Virāj was born and from Virāj Puruṣa (X.90.5ab).

Is the *puruṣa* named at the end of the verse the same as that implied at the beginning? Or can the verse be interpreted thus: from Puruṣa, the Heavenly or archetypal Man, Virāj was born, and from Virāj the earthly man, the second puruṣa being but a reflection or more concrete expression of the first? If Virāj be, as claimed by J. Gonda,[23] a creative principle representing a hypostatization of the idea of the universe, then we read in this verse that from the noumenal essence came the seed that, through the law of change, was to become the universe, which in turn gave birth to that microcosm of the macrocosm, *man*. Here is expressed in a very succinct way the law of transformation, one aspect of the dynamics of *ṛta*.

W. N. Brown sees in Virāj

> the cosmic waters from which Puruṣa arises. . . . That Puruṣa is both the son and the progenitor of Virāj is not . . . surprising: as the essence and source of all, the underlying generality, Puruṣa must precede even that out of which he rises in concrete form.[24]

We catch in this example a glimpse into one meaning of the original sacrifice of *puruṣa* as described in the famous Puruṣa hymn. This transformation of the primeval *puruṣa*, this 'birth' into more and more restricting conditions forming the various levels of the ladder of evolution is itself the perpetual sacrifice that keeps the universe going. It indirectly implies that the law of change, in the conception of the ṛṣis, becomes equivalent to the eternal sacrificial offering of the *puruṣa* by means of which the universe evolves. Both examples show the eternal rhythm at the root of the manifested universe, one principle giving birth to another which is but a permutation of itself at another level; both examples take the idea of transformation one step further than that expressed in the *sat* emerging from the *asat* and thus formulate the Ṛgvedic vision of the transformation of the ideal into the concrete.[25] Further permutations will be observed in the course of this exposition of the dynamic interplay of the gods.

The threefold division of the world

The original twofold division of the world into polarities, as envisaged in some of the Vedic hymns, is also mirrored at a more objective or lower level in Heaven and Earth, Dyaus and Pṛthivī, a twofold division which itself is further subdivided into a threefold order capable of further subdivisions: *dyaus, antarikṣa* and *pṛthivī*. Such conception of a tripartite world is found side by side with that of a twofold one and is only a more precise analysis of the component parts of the universe of which the twofold heaven and earth is a generalised view, neither excluding the other.

Heaven, Earth and the mid-or inner region, each is a *loka*, world or place of habitation for entities, whether celestial, human or sub-human. This threefold universe, so commonly referred to in the Ṛgveda,[26] is itself subdivided, each *loka* being a threefold one:

> There are three heavens; two are in the lap of Savitṛ, one is in the world of Yama (I.35.6ab).

Heaven is described as 'the threefold heavenly complex of heaven' and Savitṛ is said to

> impel the three heavens and the three earths (IV.53.5c)[27]

as well as to circumscribe the 'three realms' (*tri rajāṃsi*) or threefold mid-region. It seems evident that the three heavens refer to spheres of existence quite beyond physical visibility, since one of these belongs to the king of the dead, Yama, and since these heavens are the loka of the gods and their birthplace.[28] Within the god Varuṇa himself

> are placed the three heavens and the lower three earths disposed in sixfold order (VII.87.5ab).

Antarikṣa,[29] or the mid-region, usually translated as firmament or atmosphere, is not a physical locality such as the various translations would have us believe. The significance of this secondary level – a *loka* in itself intermediate between the purely subjective or celestial sphere (heaven), the world of luminosity, the birthplace of the gods, and the purely objective or material earth, one of the domains upon which the gods act and influence earthly phenomena – is usually altogether ignored. However, Reinhold E. G. Muller[30] shows that the original meaning of *ākāśa* is very close to *antarikṣa*. In the Upaniṣads *ākāśa* denotes the ethereal space in the heart. It is most probable that *antarikṣa* not only refers to the expanse of air or atmosphere stretching between the sky and earth, but also to an intermediate realm, a mid-region between the intangible heaven and the tangible earth, that separates the spiritual from the material, a psychological world that at the same time acts as a bridge between these, of which the ṛṣis took the firmament as the visible counterpart.

That the ṛṣis viewed the possibility of another dimension of being besides the purely physical and spiritual, as part of their vision of the universe, may be gathered from the following stanza where by implication man appears as the offspring of Heaven and Earth, the spiritual and the material:

> Heaven is my father, my begetter; here is affinity, my kinship; this mighty Earth is my mother. Here between the two wide-spreading receptacles is the womb (of birth) (I.164.33).

If we take the clause 'here between the two widespreading receptacles' literally, the word 'between' (*antar*) and the subsequent words 'womb' and 'in it' should refer to the aerial space between 'sky' and 'earth' which, literally, is not the birthplace of man. The whole connotation refers to a different dimension, a *loka* that belongs neither to the spiritual nor to the material, neither to heaven nor earth, but one which man's psychomental nature reflects. Such a *loka* may be one of the meanings of *antarikṣa*.

Pṛthivī is the material abode of which our earth seems to be one among three. Of this examples have already been given. The Śatapatha Brāhmaṇa, like the Ṛgveda, makes a clear statement:

> There are here three earths (*pṛthivī*) viz. this one, and two beyond it.[31]

J. Gonda wonders 'how these poets and priests could regard

our visible earth and the atmosphere surrounding it tripartite'.[32] Our visible earth may be tripartite, but it is also one among three, obviously three definite planets which the ṛṣis had in mind, spheres of life where beings evolve just as they do on earth, but spheres that may not be visible to physical eyes.

Among modern scholars, the most perceptive and indeed brilliant commentary ever given concerning the threefold world – *dyaus, antarikṣa, pṛthivī* – is that of Stella Kramrisch, of which certain important points are quoted here:

> Heaven and earth, as dyadic monad, are a closed unit. It must be violated, split into its components, and these must be separated lest they fall together and the world-to-be collapses. The act of creative violation and the power of keeping apart the pair so that they become Father Heaven and Mother Earth between whom all life is engendered is the test by which a creator god establishes his supremacy. He makes the Dyad into Two. He is the One and at the same time the Third, who plays the leading part in the cosmic drama.
>
> The act of separating the monad and making two who had been one is the work of a third. He is the creator god of this world. He is the demiurge who destroys for the purpose of creation; he disrupts the wholeness of the undifferentiated, separates its constituent parts, and defines their position in the new world, where men will breathe, move, and think. God Indra, in particular, is the creator of this world. He separated heaven and earth, made them two, assuming himself the part of the One and First as creator, and of the 'third' in his creative operation. As creator-destroyer he prefigures intellect in its discerning, separating, discriminating function. The self-same act of violence, of forcefully wedging wholeness apart, is the creative deed of Indra . . . as it is that of Varuṇa . . . who prefigures Indra's lordship over the gods and the world. It is the deed of the Fire as Agni Vaiśvānara, the fire that is in the sun and sunders heaven

and earth by light . . . as it is the deed of Savitṛ who separates the two by the space or air between them; it is the deed of Soma . . . and also of Bṛhaspati . . .

Three as the number of creative action and source is the number of the structure of creation of which the manifest cosmos is the base. The *tridhātu* of the visible world with its heaven, earth, and midspace is the paradigm for the trebling of this *Tridhātu;* a towering cosmic structure of three worlds, each with its Heaven, Earth, and Midspace, is envisaged.

The transition from world to world has the pattern of a chain, or triple guilloche.[33]

Indeed the three heavens and the three earths could be envisaged as globes in a chain, each globe itself being threefold.

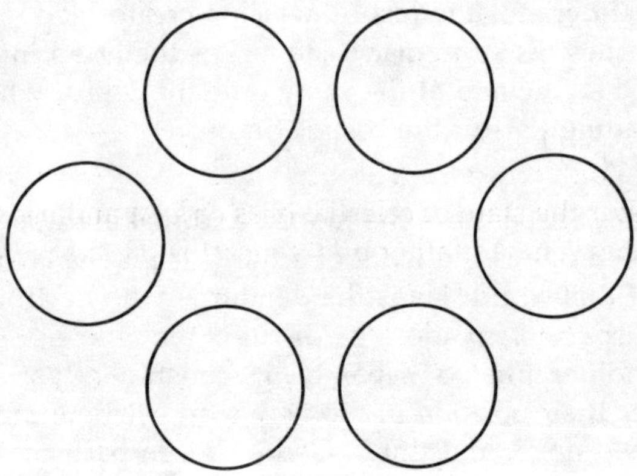

Figure 2

The gods and their cosmic interaction

Over this tripartite universe preside various categories of gods whose lower

habitations are seen, but who themselves abide in remoter, hidden dominions.[34]

Called the 'luminous sons of Dakṣa',[35] they are indeed the 'mind-born sons'[36] of that primeval differentiation which we found personified in Dakṣa and Aditi. Such an appellative hints at their inborn intelligence, skill and creative power. To describe them as 'sons' of that 'Intelligence' which is basic to all creation, without which there could be no orderly, harmonious universe, immediately suggests that they are the manifest expression of divine intelligence in action. Everywhere intelligence is seen at work, 'measuring out', 'apportioning' through time, seasons, cycles, everywhere Nature shows the handiwork of the gods who collectively could be called 'God'. 'God geometrises', as Plato once said. But it is of interest to note that the process of anthropomorphising and personifying is only incipient in the Vedas.

Heaven and Earth

At one remove from Dakṣa and Aditi, Heaven and Earth are also the heavenly parents of the gods. Heaven is either *dyaus*, the mighty father, or *svar*, that celestial realm which Indra conquers, or *vyoman*, the empyrean. Each of these words denotes a different aspect of the idea. *Dyaus* is the personified aspect, *svar* the state of celestial bliss or *loka*, and *vyoman*, the highest heaven. As father and generator, *Dyaus* is personal; as that celestial realm of luminosity and bliss, he is impersonal. Each god is capable of a threefold aspect, and sometimes more. Earth, *Pṛthivī*, the broad one, is the 'mother'. Both are compared to a bull and cow, but their personification does not go beyond parenthood. So wisdom and righteousness are attributed to them. They are almost always addressed as a pair.

Most scholars see in the fact that the gods are said to generate, prop asunder or spread out Heaven and Earth 'a downright self-contradiction',[37] or else a hopeless muddle not worth the attempt to clarify and typical of Vedic thought. This rather evidences a failure to grasp the process of permutation so notorious throughout the Vedic pantheon, and an inability to differentiate between the various levels of universal life as viewed by the ṛṣis.

Thus, Heaven and Earth are fields for development for their offspring, and so may be taken both as the 'parents' and the ever unfolding and ever renewed playground of growth, as well as those original principles of light and extension that form the basis of this manifested universe. Their offspring manifest at various levels of the ladder of creation and grow in power through the interplay of Heaven and Earth which in their very growth they themselves help to mould and transform. As principles, i.e. as the spiritual and the material fields of life, father and mother, they are the parents of all things; as specific fields of manifestation – the two worlds (*rodasī*), the two receptacles (*camvā*), the two hemispheres (*rajasī*), the two bowls (*dhiṣaṇā*) – they are realms of transformation in which the devas contribute their share in the differentiating and transforming of these very fields that give them their habitation. The use of paradox does not imply confusion. Whatever can be explained in terms of the 'below' or concrete expression applies also to the 'above' or abstract, as the former is but a reflection of the latter. So Heaven and Earth give birth to the gods and reciprocally the gods shape and further their field of manifestation into two distinct realms, and thus are said to give birth to Heaven and Earth. So

> Indra spread the wide earth out and firmly fixed it.[38]

This giving birth may be interpreted as a manifestation at different levels, hence a perpetual transformation, an aspect of the dynamics of *ṛta*. The interaction is constant, the interweaving of all forces at work within the universe being but another example of the law of transformation and balance.

As immediate parents of the gods, or *devaputra*, Heaven and Earth represent metaphysically the spiritual or 'light' principle and the substantial or 'extension' which is matter. So we find Dakṣa, ideation-intelligence-dexterity, stepped down to and reflected in that which illumines, or gives light, on the father's side, and Aditi, the infinite, the spatial deeps, stepped down to and reflected in extension or matter on the mother's side. Like Dakṣa and Aditi of which they are the permutations, they are considered by the ṛṣis as the 'ancient parents born of old'.[39] They are opposing, though not antagonistic poles of activity but rather complementary, father and mother, 'first in accordance with the *ṛta*'.[40] As they set the course of the phenomenal world on its onward march, they are called 'ancient mothers of *ṛta*', the word 'mother' being used only to denote the fact that all things in this world are fostered by them. Although still perceptible as the vast expanse of sky and earth, and personified, they are rather fields of action, of fostering growth for the transforming lives, their offspring, and certainly not 'mythological fossils'.[41] They remain throughout the Ṛgveda the primeval parents invoked for protection and for participation at the sacrifice.[42]

Given the orderly process of unfoldment and the fact that originally Heaven and Earth were considered as one, their separation – a reflection and an inevitable concomitant of the separation of Aditi and Dakṣa and of *sat* and *asat* – heralds the acceleration of the orderly unfoldment of the universe or evolution in consonance with the law of harmony:

> Thence uprose the universe; thence the spheres;
> thence heaven and earth spread out (X.149.2cd).

The gods' action is to separate or demarcate the various fields or levels of existence – the above and the below or the inner and the outer; in that sense do they shape, separate and prop asunder Heaven and Earth; so that from one homogeneous, undifferentiated whole, when the primeval parents lay fused – a state of being which the Greeks called chaos and which is described in Genesis as 'the face of the deep' (1:2) – emerged a cosmos of heterogeneity, with each division or world sphere well marked out, all in accordance with the one law of harmony which the devas express in all their activities.

To measure, sever and sustain are part of the creative act of the gods at this our space-time level. The gods' cosmic action of building or fashioning the world is often described in terms of 'measuring', the verb $\sqrt{}$*mā* expressing, in the words of J. Gonda, 'the sense of "realizing in the phenomenal world" – and this implies in three dimensional space – by applying a special technique such as "measuring" what was mentally conceived; converting an idea into dimensional actuality.'[43] Furthermore as noted by another scholar, H. Lefever, the same verb *mā*, 'measure out' or 'found', 'used in connection with the creation of heaven and earth (I.159.4) the universe (VI.47.3) and the praisesong (I.164.24) . . . is never found with *ṛta* as its object.'[44] The gods do not measure out *ṛta*, only its domain; they act in accordance with the 'blueprint' of the cosmic order conceived at the highest level and reflected at the intermediate level. For the Adityas, the sons of Aditi,

> vision the cosmic order and establish it (IV.55.2cd)

a remarkable verse which shows the gods as establishing what they have visioned and reflected in their consciousness. In his exaltation the poet sings:

> I will declare this mighty deed of magic, of glorious Varuṇa, the lord immortal, who standing in the firmament hath meted the earth out with the sun as with a measure.[45]

Varuṇa, Indra, Brahmaṇaspati, Agni, Savitṛ, are all shown as playing a vital role in the separation and propping asunder of Heaven and Earth, as is well expressed by Stella Kramrisch:

> The several deities who prop apart heaven and earth are at one in performing this feat while each expresses his own mode of creativeness. Varuṇa, the ancient and first King of gods, pillared apart heaven and earth (6.70.1). By this epiphany of his, wisdom came to the creatures (*dhīrā*; 7.86.1). But it is by his will and strength that Indra separates heaven and earth and establishes the order of his cosmos, the world in which man lives. Agni, Fire, the spark and flame of life, and Soma, the elixir of life, descend from heaven to earth and ascend from earth to heaven.[46]

Varuṇa

Varuṇa is one of the more ancient creator gods. He measures the 'ancient realm'.[47] This could refer to a former dispensation or cycle before the cycle or reign of Indra. The waters are his realm,[48] but these are not literal waters; they refer to the state of undifferentiated matter, the cosmic essence out of which the universe is shaped. Varuṇa, and his counterpart Mitra, belong to the 'earlier' (*pūrve*) gods, or to the 'first age' or 'generation' of the gods.[49] Varuṇa also separates, demarcates, measures out, possibly at a more subtle level of manifestation than that of Indra's world which appears later in the cycle of existence.

> In accordance with the law, Aditi's son, observer of the law, has spread out the world in threefold order.[50]

This could refer to the first 'measuring' of the world at the subtler level, a measuring which will subsequently be manifesting as the demarcation of heaven and earth, a task accomplished by Varuṇa and later by Indra. He gives more than just the blueprint of creation, for he sets the sun in the sky, Soma on the mountain, fire in the waters, milk in the cows and the power of intelligence in the heart (V.85.2). His cosmic action is expressive of his 'occult power' (*māyā*) and not of an original act of violence, such as Indra's rending of the dragon as a result of which he set the sun on high for all to see. Varuṇa encompasses the nights and establishes the morns. He regulates the waters, the seasons, the rivers.

Varuṇa is said to preside or rule over 'seven' – seven *lokas* or worlds? or the seven ṛṣis or streams of life? – and to pervade

> all three earths and the three highest realms of heaven; immovable is Varuṇa's seat; he rules over the seven (VIII.41.9).

> On him three heavens rest and are supported, and the three earths are there in sixfold order.[51]

He

> supports both worlds by means of the axis as the unborn supports heaven.[52]

He 'supports heaven in the seat of *ṛta*'[53] and thus presides over the cosmic order, watching over all from the very centre of all, this centre being represented not only as the root of the 'cosmic tree'

> whose root is high above (I.24.7c)

but also as that point where

> they unharness the horses of the sun (V.62.1)

i.e. that *laya* centre, or critical state wherein all manifested forms disappear from one level but to reappear at another and so on. This is the seat of cosmic order where resides Varuṇa who

> amidst the waters moves, observing men's truth and untruth[54]

He is the ocean of wisdom (VIII.41.8), he is clothed by the waters that shower their sweetness.[55]

The identification of the waters with the 'womb of *ṛta*' may seem strange at first sight, but if we keep in mind that the cosmic waters are the 'essence' of primordial matter, the later *ākāśa* of Indian philosophy out of which the universe is moulded in a series of more and more crystallised aspects, then there should be no difficulty; the expression points once more to the fact that the law of order, and of evolution, of unfoldment, of adjustment, manifests at the very origin of time, in the cosmic waters which are so to speak its very womb and which as cosmic essence, or 'atoms' as we might say nowadays, develop, assume more and more concrete appearances in accordance with the 'law' of change! Varuṇa, as originally conceived by the ṛṣis, is the very soul of this cosmic essence, the Overseer and Orderer who watches over Creation, and, as the embodiment of the very Law that is at the root of the universe, becomes for early Vedic man the 'far sighted', 'thousand eyed',[56] Lord of cosmic order and its reflection in human society, viz. ethics. He becomes the moral god par excellence.

In the later Hindu mythology Varuṇa loses his lofty qualities and overlordship of the Cosmic Order and becomes a god of waters and darkness. This degradation is already visible in the Atharvaveda and the Brāhmaṇas.

Indra

Perhaps the most insistent description of godly creative action centres on Indra, who was the most popular of the gods. He repeatedly makes the '*sat*' out of the '*asat*', separates, props asunder heaven and earth. His cosmic action,[57] described as making firm the quaking earth, more widely extending the middle region and supporting the heavens, might hint at the physical formation of our world, the original convulsions to which the globe was subject, Indra personifying the ordering intelligence behind the whole configuration of the cosmos in movement.

We notice the extension of the mid-region which is not only the atmosphere as such, the space that lies between the earth and the vault of the sky, but also that which separates and links one pole of manifestation and the other, the spiritual and subjective symbolised in heaven, and the phenomenal and objective symbolised in earth, which originally were in close 'embrace' born together from the same Source. From the psychological standpoint this means the unfolding of an inner world between the spiritual and the physical or objective, a mid-region which, at the human level, is equivalent to the psychological world typical of man which separates and at the same time provides a bridge between the spiritual and the physical.

Indra is not the mere storm god which later Indian lore made him into, but his association with the sun and the finding of light (III.39.7a) and certain of his feats point to his fundamental psychological importance. He is the space-maker, an expression of multi-levelled meaning. He is the light-bringer whose action of severing the darkness, of slaying Vṛtra, dragon of chaos, makes him the personification of mind's struggle for freedom. Indra modified the primeval state of undifferentiation into a differentiated world, with distinctive realms, sequence, time, contrasts, light and darkness. Mind's function is to give shape to chaos, to make 'pathways', distinctions, demarcations:

> With the sun's help he made pathways throughout the darkness that extended pathless (VI.21.3ab).

He is even identified with Sūrya:

> Sūrya is he . . . he has destroyed with light the black-hued darkness (X.89.2).

and therefore is considered the 'illuminer of nights' and 'generator of the sun'.[58] He is the 'winner of the light', or 'heaven' and of the 'god-like waters' (III.34.8).

> Through might and conflict he made space for the gods (III.34.7)

i.e. gave freedom to the gods, space in the Ṛgveda being often equivalent to freedom. The gods and Heaven and Earth engendered him to crush all obstacles (III.49.1) and to him they gave *asura* power, that is, the creative power, for having achieved the goal for which he came into being.[59] He is three times compared to a ram, and in I.52.1. he is

> the ram that finds the light of heaven.

According to A. A. Macdonell:

> Regarded as a whole the attributes of Indra are chiefly those of physical superiority and of dominion over the physical world. Energetic action is characteristic of him . . . Indra is a universal monarch, not as the applier of the eternal laws of the universe nor as a moral ruler but as an irresistible warrior whose mighty arm wins victory, whose inexhaustible liberality bestows the highest good on mankind.[60]

As already seen there is more to Indra than just physical superiority. Indra, in the Ṛgveda, is the expression of mental vigour, alertness as well as bravery, courage and so on. He is the 'thought-bestirrer' (*codayanmati* VIII.46.19), prayer makes him increase, or wax stronger.

> Thou becomest great by virtue of the sacred word.[61]

He storms heaven through the kindling of his inner fire (X.167.1d) and is said to be the 'first to possess mind' (*prathamo manasvān* II.12.1), that mind that is the separator, the active organising factor both in nature and in man, which is instrumental in bringing the light; it rends the darkness, hence the image of the lightning. He, the 'meditating god' (X.32.1), discovers the nectar of immortality in the third sphere of heaven:

> In the threefold heaven, in the triple splendour, he found the hidden immortality.[62]

As Indra drank the Soma the sun arose in the heavens (IX.86.22); in the rapture of Soma he set free the cows held fast within the rock (VI.43.3), loosened the rock and 'made the cow easy to find for prayer' (X.112.8). The unconscious is the rock in which is hidden the treasure that yields enlightenment, the milk or nectar of immortality. Matter is also the matrix in which is concealed the nourishing substance of both poles of manifestation: spirit and matter. The prize Indra wins is thus not only the sun 'that men may see' (IV.16.4), not only the waters of life or the 'seven rivers' that are released through his action (I.32.12), but also Soma who becomes his beverage *par excellence*. Soma is even equated with his thunderbolt.[63] The lightning flash illuminates the mind and in that illumination is experienced the state of immortality, of wholeness wherein all dimensions are merged in one supreme oneness.

> Soma, Indra's juice, becomes a thousand-winning thunderbolt.[64]

Indra is the fighter in man, the liberating mind that forges its path amidst the darkness of the unconscious, shapes it into an ordered world, and opens out the way to the 'divine' hidden deep within, in the 'third sphere of heaven'.

Man invokes him in the 'light-bestowing conflict'

for through 'conflict' (*yudhā*) Indra found for man the 'way to fulfilment' (X.49.9).[65]

Brahmaṇaspati

Brahmaṇaspati, whom Stella Kramrisch describes as 'the lord of the inner upsurge in hymnal, mantric inspiration', also assumes the role of a demiurge or creator god. The idea of blowing, or whirling, or welding the worlds finds expression in one of the creation hymns where we have the picture of the Lord of prayer, Brahmaṇaspati, welding the worlds into shape as a blacksmith:

> Brahmaṇaspati welded these (the two worlds) together like a blacksmith (X.72.2ab).

The verb √*dham* to blow, exhale, kindle, melt, forge, and the word *karmāra*, artisan, blacksmith, point to the action of fire as it fashions the worlds. God is the artisan, the blacksmith who smelts his universe into shape.

At the beginning of all and at every level, was *fire*, whether expressed as *tapas*, the divine contemplation blazing to incandescent manifestation, or as Agni, the Lord of Flame, or as *kāma*, desire, primordial seed of consciousness that sparked off creation.

Agni and Brahmaṇaspati are vague personifications of this fire, for their action is intelligent and creative, and Brahmaṇaspati, in many instances, is but another aspect of Agni. He forges Heaven and Earth but also waxes in grandeur through the very parents that he moulds, he himself being called the 'father of the gods'. But there is a peculiarity about Brahmaṇaspati that distinguishes him from the other creator gods. He embodies the divine *word*, *brahman*, which is of the essence of the devas. We are told that, by virtue of the sacred word, (*brahman*) Indra waxes great (X.50.4). That sacred word, or primordial vibration, forged the worlds, is at the basis of all and is harboured in the heart of the ṛṣis. Brahmaṇaspati embodying it is thus the 'father of the gods', although

Heaven and Earth in another sense are also his parents. Brahmaṇaspati

> mightily established by means of sound, the earthly limits as the threefold world (IV.50.1).

The primeval vibration, the sound divine, acts like fire in shaping the universe; hence the blend of two ideas, that of *sounding* and that of *kindling* as the function of the one god, which to us appear completely different.

Agni

The primordial creative urge from which proceeds 'this whole universe', called 'desire' in the *nāsadīya* hymn of creation, the catalyst that brings about the change from one quiescent state to another, dynamic state of being, may be viewed as personified in Agni, 'son of the waters', 'embryo of the waters',[66] the 'first born of *ṛta*' at the beginning, in Aditi's bosom, in Dakṣa's birthplace,[67] that is, manifested in accordance with the law. In Agni is exemplified the birth or appearance of the same principle at different levels of manifestation. For he is said to be created by Heaven and Earth, by the Waters, by Tvaṣṭṛ (X.2.7), but at the same time he appears right at the beginning of the aeons as shown above, for *tapas* is an expression of cosmic creative fire. It is the process of raising this fire. This idea of births from various levels and centres is repeated in many verses. Thus:

> From heaven first came to birth Agni; secondly from us came the knower of births; thirdly the manly-souled (appeared) from within the waters.[68]

He exists at all levels, is the basis of all, hence appears in the waters, in the stones, in the herbs, is 'head of heaven, earth's centre', 'engendered by the gods' yet 'father of gods' (I.69.1) and their giver of immortality. He is common to all men (*vaiśvānara*), their very centre (*nābhiḥ*), like a pillar supporting them, immortal guest midst mortal bodies. Men find him crouched in the cave of the heart when they fashion their mantras there (I.67.2). Primordial spark, he is also the youngest of the gods and born of Tvaṣṭṛ, the artisan of nature. Such a multiple birth shows the one creative and destructive principle at work on the various rungs of the ladder of creation.[69]

Stella Kramrisch sums him up thus:

> The three main forms of Agni, in the process of creation, are: Apām Napāt in the flood of the Uncreate, before and beyond creation; Vaiśvānara in Heaven, the light that shines for all men, and Jātavedas on earth, the Knower of beings. This great ternary of Agni in the process of creation is accompanied by the triple manifestation of Fire in the visible cosmos, the sun in heaven, lightning in the air, fire on earth. The great ternary of Agni in creation is ontological and cosmogonic; his ternary in manifestation is in the triple world of Indra. Indra and Agni are twin brothers (6.59.2).[70]

Tvaṣṭṛ

At the level of what we know as nature, we find Tvaṣṭṛ the divine artisan, the great, skilful fashioner, he who, according to the Atharvaveda, is 'in the beginning the counterpart of the waters' (IX.4.2). He might have represented originally the male creative power inherent in the waters, the female enfolding substance. He is called the one born at the beginning, *agraja* or *agriya*.[71] Universal stimulator, he receives that epithet *savitṛ* which was to be attached to the solar deity; he is called the skilful-handed one (*supāni*), the omniform, the 'faithful to the law' (*ṛtāvan*) who combines both the male and the female generative power. Hence the Atharvaveda calls him 'the pregnant male, stout, filled with milk' (IX.4.3). Presiding over birth, he places the life seed in creatures, shapes all forms, adorns Heaven and Earth with variegated forms, gives nature her final shape, forges the thunderbolt of Indra, fashions the soma cup for the gods and sharpens his axe for Brahmaṇaspati. The axe of time, Tvaṣṭṛ's attribute, is ever at work in fashioning and refashioning nature's myriad forms in accordance with the law (*ṛtāvan*). The whole of nature's creative and varied activity is compressed in the two epithets used for Tvaṣṭṛ: 'the omniform, vivifying deity'.

It is noteworthy to remark that these originally descriptive epithets of one and the same deity in due time were separated from their owner and addressed as a personality in its own right. Thus the stimulator (*savitā*) side of Tvaṣṭṛ became related to the sun as the vivifying power, Savitṛ, to whom the great *gāyatrī* mantra is addressed, whilst the omniform aspect became differentiated and personified in a monstrous son of Tvaṣṭṛ, Viśvarūpa. Furthermore other gods such as Parjanya, Prajāpati, Viśvakarman take on the cosmogonic functions of Tvaṣṭṛ so that one may suspect that all are but various names describing the same creative power of nature, or facets of the one power.

Agni, Brahmaṇaspati, Tvaṣṭṛ are instances of that constant permutation and overlapping in function so characteristic of the gods as well as of their actions at the various levels of manifestation. Herein lies the reason why it is difficult to assign each god to a particular group as the naturalist classification tried to do. The 'all-pervasiveness' of Agni makes him difficult to treat as a purely earthly deity. The same applies to Brahmaṇaspati and Soma, both of whom are at the 'beginning', in the heavens, and on earth.

FIVE

The working of the Law

The laying down of the cosmic law

Creation, at the secondary level, is, as we have seen, a question of measuring, ordering, establishing, instituting; a process of organising, fashioning, a laying down in the right order according to certain norms, ways, rules, ordinances, rhythms, inherent in the *ṛta,* regulating the unfoldment, development, movement of all things; hence creation in its secondary sense is an 'establishing', this being the function of the gods.

Three words, in the Ṛgveda, express this establishment, maintenance and working of the law: *dhāman, dharman, vrata*.

The verb 'to establish' (√*dhā*) represents the foundation in full accordance with 'law'; it is the laying down of the cosmic order in the space-time world in strict accordance with the rules which stem from, indeed are the expressions of, the fundamental order, the inherent harmony at the core of the universe, the blueprint of *ṛta*.

The Adityas first vision the cosmic order – they are called '*ṛta* visioned' (*ṛtadhitayaḥ*) – as the pattern which man's poetic imagination has described as 'eternal in the heavens', then establish it at the phenomenal level.

> They have established the seasons, months, days, nights, the worship and the holyword (VII.66.11ab).

The sacrificial ritual is here brought in line with time's sequence, and given the seal of godly statute.

The specific sphere or domain of activity of each god within the vaster domain of *ṛta* is called *dhāman*.[1] In the '*dhāman* of *ṛta* the gods rejoice' (IV.7.7b). This is the establishment of manifestations of *ṛta*. *Dhāman*, as J. Gonda explains, connotes 'locations of modalities or manifestations of divine powers'.[2] A glimpse is given us into a primordial order wherein Varuṇa 'established his dominion of old'.[3] Since Varuṇa's sphere of influence extends over the whole phenomenal world, the world is his *dhāman*, his peculiar domain. Indra claims that he has 'established many a sphere of *ṛta*'. Agni 'marks out' the many 'manifestations of *ṛta*'. He 'circumscribes seven dhāma',[4] seven seats or special domains wherein the law is fully expressed, possibly the sacrificial seats wherein is enacted the drama of the cosmic order, or the worlds within worlds of the Vedic vision which, as we have seen, he pervades from the innermost to the outermost.

Each *dhāman* is a sphere of harmony established in and working with the overall harmony of the *ṛta* at the innermost core of which is found the secret of immortality.[5]

The second law word, *dharman*, from √*dhṛ*, to buttress, support, which often governs *ṛta*, seems to have meant in Vedic times that which holds together through the inherent law of harmony, cohesion, that which maintains the foundations of the law. *Dharman* is a more specialised aspect of *ṛta* and is used in the sense of specific statutes that sustain, regulate and order the course of the universe; the term *ṛta* referring to the whole, i.e. the cosmic and social order, the truth, the sacrificial rite, the term *dharman* to specific regulations binding under that whole. The fixed norms of cosmic order manifest themselves as the divine statutes which uphold the march of all things and maintain the universe in every aspect.

One of the gods' proper functions is thus the 'maintenance' of the *ṛta* (*ṛtasya dharman*):

For the sake of upholding and supporting the world,

> through (Agni's) own power have the gods expanded (X.88.1cd)

their action standing

> firmly fixed by statute and ordinance (V.72.2a).

Through the same ordinance, or what is laid out as the norm,

> the refulgent sun journeys, a god amidst the goddesses (I.160.1d)

i.e. the sun travels between Heaven and Earth, here taken as the mothers, in accordance with the laws determined by the cosmic order. Accordant with the same *dharman*, Indra

> establishes the rivers in their course (II.13.7a)

each god following the law proper to his being, in other words his own *dharman*.

Several gods are intimately connected with the 'foundation of truth': thus Agni, Savitṛ, Mitra-Varuṇa. Agni's truth is the good that he does to the worshipper; he inspires man, he uplifts him and creates the link between god and man, he is the harbinger of divine manifestations. His is the task of the 'overseer of the *dharman*' in his quality of custodian of the ritual as commemorative of the universal rite of creation. Savitṛ, like Varuṇa, is he whose *dharman* is true (*satya-dharman*), i.e. accomplished, he who maintains the statutes. But Varuṇa, whose insight detects all, the good and the evil, what has been and what will be, whose eye is the sun, is the custodian of the *ṛta* as a whole, he is the very embodiment of truth.

Truth, in its metaphysical sense, at this level of the gods, is thus the realisation or implementation of what has been conceived as a pattern at an even more abstract level and is 'visioned' by the gods ere they express it through their action. It is the divine harmony made visible in all the workings of nature. The world order is the mirror of the harmony that exists between the essence of being (*sat*) and its outer activity. This is truth. Thus a thing will grow only in accordance with what is inherent in its seed; the acorn will always produce an oak tree, a human embryo a human being; when ignited under proper conditions, fire never fails to be kindled, for we have touched its very nature. All natural activities are visible effects of an inner harmony which constitutes the supreme Order. This is the truth; this is the gods' truth of action, of being. This is truth in the Cosmic Order. Truth itself is said to be

> established on the unshakeable foundation of heaven.[6]

The accomplishment of the law as an expression of truth, *satya-dharman*, manifests in human society as social order and right human relationships. So the rules laid down for human behaviour are best calculated to promote social harmony and prevent society from breaking down. *Satya* truth as related to society becomes *dharma*. The subsequent development of *dharman*, which as *dharma* came more and more into usage, made the word *ṛta* fall into disuse.[7]

The third law term, *vrata*,[8] has in the Ṛgveda a different connotation from its later meaning of 'vow'. Like *dhāma*, it also refers to the sphere of divine influence or the right 'course of action', the particular way according to which each god performs his function, hence statute, norm. Each god in his own characteristic way upholds his own mode of action, his own *vrata*, all of which vratas form the established norms, the divine 'inroads'. By means of these Varuṇa, for example, maintains order in the world, these being expressions of the over-ruling *ṛta*. So, of their own nature, acting in perfect harmony with the whole,

> 'the gods follow the ordained-ways of *ṛta*.[9]

The world

> Heaven and Earth, the waters, the plants and the trees, did the holy ones engender following the ordinances. In graciousness, they filled the midregion with heavenly light (X.66.9abc).

This heavenly light is not just light (*jyotis*), but *svar*, the luminous world of heaven which in the later Hindu scriptures becomes *svarga*, the heaven of the blessed.[10]

Neither Varuṇa, nor Indra, nor any other god infringes the *vratas* of Savitṛ:

> In accordance with *ṛta* godly Savitṛ toils; he extends far and wide the peak of *ṛta*.[11]
>
> The wild beasts spread through desert places seeking their watery share which thou hast set in waters. The woods are given to the birds. These statutes of the god Savitṛ none disobeyeth.
>
> With utmost speed, in restless haste at sunset Varuṇa seeks his watery habitation. Then seeks each bird his nest, each beast his lodging. In due place Savitṛ hath set each creature.
>
> Him whose high law neither Varuṇa nor Indra, nor Mitra, nor Aryaman, nor Rudra breaketh, nor evil-hearted fiends, here for my welfare him I invoke, god Savitṛ for worship (II.38.7–9 Griffith's translation).

The waters obey him, night comes at his bidding, he wakens all, stretching out his arms to protect all. Each kingdom of nature, each species, has its own domain and its own activity under his guardianship. The god fills the whole of Heaven and Earth and makes the 'holy song' (*ślokam*) which he

> fashions for the sake of his own dharman (IV.53.3).

The holy song may be taken as that which comes to the lips of the poet inspired by the solar splendour, but it may also be taken in a different sense: the universe is the song of the solar god, that song which he cannot help but sing as he performs his varied tasks. This is confirmed in a verse which tells us that 'the gods revel in the sacred song of *ṛta*' (*sāman* I.147.1d).

> To the god Savitṛ who makes the *ṛta*, to Indra the serpent slayer, flow the waters; day after day spreads out the sheen of the waters; how long has (time) passed since these waters were first outpoured (II.30.1).

Savitṛ in making the *ṛta* is represented as responsible for the orderly recurrence of natural phenomena. The paths along which all things move are fixed, hence with 'fixed decree' (*dhṛtavrataḥ*) he rules his mighty course (IV.53.4). These *vratas* are also described as triple and therefore apply to the three heavens and the three earths of Vedic cosmogony:

> Thrice does Savitṛ circumscribe with his might the mid-region, the three realms and the three luminous spheres, impelling the three heavens and the three earths; indeed he protects us with his triple statute (IV.53.5).

Dawn, the

> ancient, youthful, prolific goddess, the all bounteous, follows along the established norm (*anu vratam*).[12]

As the dawns perpetually ascend the sky and pervade the firmament

> they generate the divine inroads.[13]

In obeying their own intrinsic law they show us the path we should follow, those intangible inroads whereby we learn to harmonise ourselves with the whole.

Similarly for those who follow Indra's *vrata* Heaven and Earth and the mountains stand firm. Varuṇa's *vratas* yield in importance to none of all the other gods for the gods themselves are said to follow them:

> All the gods follow the *vrata* of Varuṇa.[14]

Varuṇa, the *asura*, the lord of righteousness (*ṛtayan*), the guardian of *ṛta* whose 'statutes are firmly established', seems to have been recognised at least at one time as the supreme lord of all, 'ruler of this world's manifestation'. He has *ṛta* as his 'form'[15] and figures as the god in whom *ṛta* seems to be incarnate; his is the dynamic expression of *ṛta*. Such

> statutes (*vrata*), unmovable, are firmly-fixed in thee as on a mountain (II.28.8cd).

In some hymns he steps into the role of the all-creator in whom all the worlds abide, in others that of the all-disposer of justice to whom the sinner turns in repentance. He is the mighty upholder of order – physical, as he controls natural phenomena, and moral as he

> perceives man's truth and untruth

and beyond both as he

> knows all things transcendental and beholds what has been and what will be.[16]

So when we read of Savitṛ making the *ṛta*, of the rivers running in accordance with Varuṇa's *ṛta*, we realise that what is meant here is the particular sphere of action over which the god has jurisdiction which, taken together, form the great *ṛta*, Varuṇa seemingly exercising lordship over the whole sphere of *ṛta* in which the *vratas* are fixed as on a mountain.

Three of the law terms[17] are brought together in a verse that sums up the action of some of the Adityas:

> By means of the *dharman*, with the magic power (*māyā*) of the asura, O wise Mitra Varuṇa, ye protect the *vratas*; in accordance with *ṛta* ye rule over the whole world (V.63.7).

The *māyā* or occult power of the gods (a power which shall be investigated in the next but one section) is here practically an expression of the law as wielded by the gods through which the pathways (*vrata*) of the law are protected. Thus the various ways (or statutes) of upholding (*dharman*) the Law are the various modes of that *ṛta* whose custodians the Adityas are. The law itself acts as the very rhythm of becoming, the undercurrent of that transformation according to which the universe proceeds in its endless evolution. Hence the law, *ṛta*, is in one sense apart from and beyond the gods, and in another sense their very being and mode of action so that they are occasionally said to 'make the *ṛta*'. It exists at a transcendental level but manifests through them on every rung of the ladder of existence. They are its intelligent, dynamic agents.

Impersonal law and morally perfect gods

The gods thus stand on the constructive, evolving, expanding, forward directed, orderly side of the cosmic process. They are responsible for the divine impulse (*deva-hitim*). They are thereby the custodians of the whole of creation, its regents, its builders, its 'disposers' (*dhātṛ*). For this reason they are considered 'offsprings of *ṛta*' (*ṛtajātāḥ*), said to be the 'first born of *ṛta*', 'dwellers in the seat of *ṛta*', in that realm where the law is perfectly manifested. They are born and wax strong in accordance with the *ṛta*, i.e. as a result of observing and serving it. They thus increase in and foster the *ṛta*:

> These, Aditi's sons, undeceivable, have grown mighty in the abode of *ṛta* (VII.60.5cd).

They have taken up their stand on the side of order and to this effect first conquered all kinds of disorders. In man's eyes they are sinless (*anāgas*) because they conform to the *ṛta* (*ṛtavarī*) and are themselves considered the 'great *ṛta*' (*ṛtam bṛhat*) i.e. the very embodiment of the law they serve. They obey the *ṛta* since they cannot act against their own nature and at the same time they are its 'knowers', its 'charioteers', its 'guardians' and its 'lords', the perfect expression of that which manifests through them, by them and with them:

> Law abiding, born in law, sublime fosterers of law, haters of falsehood (VII.66.13ab).

> herdsmen of the supreme-law, whose decrees are truth (V.63.1ab).

Therefore they

> rule over the whole world in accordance with the *ṛta* (V.63.7c).

> serving the *ṛta*, by means of *ṛta* they have attained to powerful might.[18]

They are '*ṛta*-visioned' (*ṛtadhītayaḥ*), cherishers of the *ṛta* (*ṛtaspṛsaḥ'*) and therefore 'true' (*satyaḥ*). In such guise they shine forth and work their wonders, the latter being the marvels of nature's phenomena, her predictable cycles and profusion of beauteous forms.

The ṛṣis conceived an impersonal eternal law to which all, even the most highly evolved beings such as the gods, are subservient. All the statutes (*dharman*) and ordinances (*vrata*) of the gods, as perceptively observed by H. Lefever, 'are the expressions of a transcendent, objective law which . . . the gods "find" rather than create . . . the gods do not create ṛta . . . they find ṛta – something existing apart from and above them' – ṛta being in the estimation of the same author the 'cosmic and dynamic principle of Law and Order', 'the transcendent objective law of the universe, applying not only to ritual and natural phenomena, but also to human conduct.'[19]

The gods guard the statutes of the one law both as protectors (*ṛtasya gopāḥ*) and as 'debt-exactors' (*cayamānāḥ – ṛṇāni*), their very action, or ways of acting constituting those statutes, hence they are said to follow the ordained course of *ṛta*. The mention of 'ropeless bonds' (VII.84.2ab) whereby Varuṇa binds the sinner is significant enough as showing a universe governed by laws which bind its various parts together as one unity under the regency of Mitra-Varuṇa and the other Adityas.

The issue of a moral order at the godly level or in what we might call the cosmic vision of the ṛṣis is taken up by Balbir Singh Gauchwal in his article on 'The metaphysical foundations of Hindu ethics and religion'.[20] He asks 'whether the moral law in its metaphysical aspect can operate by itself or requires some conscious personal agency to manifest and operate it.' This idea is based upon the assumption that the Vedic thinkers unanimously formulated, 'an objective moral law constituting the eternal moral order of the universe – called *ṛta, dharma* or *apūrva*'.[21] To think in terms of a cosmic moral order is to bring in a purely human dimension at a level where the purely human is bypassed. The objective moral order of the universe exists solely in man's mind. Its counterpart in the universe is harmony, equilibrium. The moral law, as formulated by the Vedic ṛṣis and applying to humanity, is only one aspect of the universal law as man envisages it, one aspect that in its specific relation to him he considers as moral, but not the whole of the law. That whole could be more appropriately summed up, not as 'the objective law of goodness', but simply as the 'law of harmony' of which the gods are the active, intelligent exponents, hence the law of the solidarity of all beings which at the human level is reflected in the human values of love, togetherness, helpfulness, service, etc. That which is consonant with the overall harmony will, in the human sphere of activity, be considered moral or good, hence the norms of social as well as personal ethics that form the basis of all civilisations.

In this respect there is no need of a supreme God or Arbiter of morals, except in so far as a personalised representative of the one Law to whom the human being renders account of his deeds or misdeeds, and begs for forgiveness, in whom he can place his trust, hope and love, seems to be more comprehensible to the average human mind, than an impersonal law of which only one aspect touches man in his most vital activity, i.e. living in righteousness, distinguishing between good and evil. This accounts for the ingrained prejudice which nineteenth century scholars all evidence when faced with the interpretation of the law of karma or cause and effect. It is perhaps for this reason of average human inability to understand an impersonal law, that we do find one Vedic God standing out prominently as the 'Guardian' *par excellence* of *ṛta*, one to whom appeals can be made and whose forgiveness can be invoked, one who is closest to the Christian and Judaic idea of a personal God. This is Varuṇa, the Lord of Cosmic Order.

Of the unnamed God of Ṛgv. X.121, the poet says:

> [He who] through his truth-accordant-decrees (*satya-dharma*) gave birth to heaven, who is earth's procreator, cannot harm us (X.121.9ab).

The law that is the foundation of all is truth and all-binding. Truth at the cosmic level means the working out of all things in harmony with their innermost core, the outer expression of the inner being, hence harmonious interlinkedness, for being is ultimately one. As creation proceeds in accordance with the law inherent to its very core, all creatures must follow the law, i.e. live in accordance with it. Minerals, plants, animals live thus. The animals, however wild, have their own code of behaviour which none infringes. Only in standing against the law can one be harmed, and, generally, only man again and again stands against it. Superficially the above verse may be thought in contradiction with certain prayers addressed in particular to Varuṇa the God whose noose ensnares the sinner. But the meaning is that man is harmed only in so far as he transgresses the law, thereby laying himself open to such specific harm as is brought about through sin; a man in harmony with the Law which is Truth, and therefore in harmony with himself, his own deeper self, there where his link with the highest or innermost is at its strongest, cannot be harmed, harm being contingent upon man's action and nothing else. Hence God cannot harm him.

> Sweet blows the breeze for one who lives by truth,
> rivers pour for him sweets (I.90.6).

This whole aspect of the divine law and its tacit mandate unto man, and man's recognition of it and attempt to live up to it, as expressed in the Ṛgveda, is quite in harmony with the ideas found in the magnificent Psalm 119 of the Old Testament. There, the Psalmist prays to be kept ever on the divine path. 'O teach me thy statutes' is like a refrain spelling out man's heart desire to follow righteousness.

> 'make me to understand the way of thy precepts; so shall I talk of thy wondrous works' (27)

> 'And I will walk at liberty; for I seek thy precepts' (45)

In keeping the law there is freedom; this freedom man experiences when he is at one with his inner self and therefore abides in harmony with the All. This is the very core of the Vedic experience of the cosmos: wholeness (*sarvatāti*).

> 'We choose freedom and wholeness', cries the bard of Ṛgveda (X.100).

The Psalmist goes on:

> 'Thy statutes have been my songs in the house of my pilgrimage' (54)

Here again we find a point of contact. For the Vedic bards the Law is the song of the cosmos, the cosmos expresses itself through the Law, which is harmony, cosmic order, and so

> 'the gods revel in the sacred-song of *ṛta*' (I.147.1d).

Opposition and ambivalence: *asura* and *māyā* in the Ṛgveda

In a universe based upon the subtle balancing of opposite tensions, it is inevitable that even at the godly level the two great forces which we call constructive and destructive, evolving and involving, expansive and restrictive, centrifugal and centripetal make themselves felt as indeed they make the web of the universe. Because of this, the gods encounter opposition in their constructive work of laying down the laws and demarcating the realms of the universe. These two opposing factors were expressed by the ṛṣis in the enmity between certain categories of gods, but the tension and conflict between these two is left somewhat blurred in the Ṛgveda. In later Hindu mythology it becomes the never ending struggle between the gods and the asuras and the degradation of both to our human level of jealousies, quarrellings, fighting and deceit.

The word *asura* in the Ṛgveda does not bear the connotation of demon as it assumes later. *Asu* means breath or life-energy (+ *ra* from √rā to possess and also to grant). *Asura* thus not only means the possessor of the life-breath but also its bestower. We may surmise that what differentiates the *asuras* from the *devas* in the Ṛgveda is precisely their capacity to create; they are the creative gods; in as much as a *deva* rises to that accomplishment he is also entitled to be addressed as *asura*.[22] The title *asura* was, in Ṛgvedic times, applicable to highly spiritual entities, both for their creative and ruling capacities and especially for the magic power which they could wield, and not to demons only, as in post-Vedic mythology. In fact the highest devas are *asuras*. Varuṇa, Savitṛ, Indra are each described as 'wise *asura*'; Dyaus, Agni, Brahmaṇaspati are also given that title. *Asura* power is granted to Indra because of mighty deeds:

> Even as the power of Dyaus, to thee O Indra, all asura sway was by the gods entrusted, when thou, impetuous, leagued with Viṣṇu, slewest Vṛtra, the dragon who enclosed the waters.[23]

But among the class of the *asuras*, there are already mentioned certain names that belong to the dark side of the cosmos. Certain *asuras* are opposed to the *Adityas*, the sons of *Aditi*, the Boundless, the Infinite, and Indra is shown as fighting them. In the Adityas we discern the principle of expansion, freedom, light, the forces of evolution, constructing and supervising a world where law reigns supreme. In the opposing forces, the Dānavas, which will become the *asuras* of later meaning, we notice the principle of constriction, bondage. Thus *Dānu* (bondage, restraint) is the mother of *Vṛtra*, the constrictor, and of the seven *Dānavas*.[24] The seven *Adityas* of early Ṛgvedic days stand so to speak facing the seven *Dānavas*. But of the latter only Vṛtra, the dragon of obstruction, receives full attention on the part of the bards. In him is embodied chaos, darkness, opposition, constriction. Indra fights him but he is also said to fight 'all the furious gods' (I.30.3 and 5), though no explanation is given.

The term *asura* in the later sense of demon is mentioned but a few times in the later portions of the Ṛgveda.[25] Out of these few vague and scattered hints, there emerges an indefinite, yet sufficiently outlined picture of the forces of light, construction, evolution, personified in the Adityas, and the forces of darkness, constriction, retrogression, personified in the Dānavas and others, all originally belonging to the category of *asura* and wielding the same power, *māyā*.

The conflict is thus originally between the devas as *asuras* and the *daityas* or sons of Diti or Dānu, between the Adityas or sons of 'Freedom' or the 'releasers', and the sons of 'Constriction' or the 'restrainers'. In the Śatapatha Brāhmaṇa (1.2.4 ff.) these opposing forces are named the sons of the same Father, Prajāpati, the Lord of Beings, but of different Mothers, Aditi and Dānu, the opposing powers of release and constriction.

The primordial gods born in the *asat*, the undifferentiated state prior to the differentiation that became the cosmos, would have to embody the self-assertive principle of consciousness capable of breaking through the status quo, of piercing through the state of darkness or undifferentiation. The later scriptures, the Śatapatha Brāhmaṇa (2.4.2.5) and the Taittirīya Saṃhita (1.5.9) associate darkness or night with the asuras and day with the devas. This becomes a matter of fact in the Viṣṇu Purāṇa and other Purāṇas.

Self-assertion as separative drive is essential to manifestation. The action of the asuras of the Vedas is the action of the differentiating, divisive intelligence that strove to assert itself at the beginning of the aeons or during the first part of the cycle, to establish its self-consciousness, to live its separate existence, to make cosmos out of chaos, to make a world out of the undifferentiated state of darkness that prevailed, as described in the famous Ṛgvedic hymn of creation (X.129). All the great asuras are examples of this self-assertion of high intelligence breaking through the circle of darkness and demarcating the various realms of the manifested world. The most popular figure in this respect is Indra, the god with the hundred powers (*śatakratu*); he is the active, martial sovereign who asserts his own separative existence by issuing from his 'mother' in the way he pleases, by rending asunder the dragon of matter, by making cosmos out of chaos (the *sat* out of the *asat*, VI.24.5b), by setting the sun on high for all to see, releasing the rivers, and so on. He is twice called asura (I.174.1 and VIII.79.6).

The fall of the asuras from the high creative gods that they originally were in the Ṛgveda to the 'antigods' of the Purāṇas, might imply the idea that certain asuras allowed the self-assertive force which originally had propelled the spiritual into the material to dominate their lives till it became complete separativeness with all that this entails. This attitude is symbolised in the story told in the Śatapatha Brāhmaṇa (5.1.1 ff.) that the asuras 'even through arrogance went on offering into their own mouth', whilst the gods 'went on making offerings unto one another', as a result of which Prajāpati gave himself up to the gods.

Asuraship seems thus, in the Ṛgveda, to refer to a certain status, a status of attainment characterised by a power; the *asura*, on the constructive side of nature, creates, orders, organises and his fundamental means of creation is described as *māyā*, a word whose meaning is difficult to pinpoint exactly and which, for this reason, has aroused a great deal of controversy.

Māyā[26] as a creative power is not exclusively the asura's possession, as the devas as well as those who were in due course to be deified (e.g. the Ṛbhus) also possessed it, but it may be theirs in the highest degree. This skill is also the prerogative of demons, the asuras on the destructive side of nature. The *māyāvin* can be either godly or demonic.

Asuramāyā, the term used in the Atharvaveda (III.9.4) refers to divine creative power as well as to that power of transformation peculiar to gods and demons, later to be used by the demonic powers for deceptive purposes.

K. S. Varma points out that *māyā* is used in three different senses in the Ṛgveda:

1 it means the supernatural power of gods – especially Indra, Mitra and Varuṇa – to transform themselves or assume strange forms (cf. Ṛgv. VI.47.18: 'through his power of *māyā* Indra wanders in many forms; harnessed are his thousand horses');
2 it is often used to indicate the world sustaining power (cf. Ṛgv. III.38.7; IX.18.3, etc.);
3 it means the deception or cunning exercised by the asuras (demons) in their fight against the devas (gods). Like gods, demons also have power of 'assuming any

shape at will, but they excell the gods in the power of subtle trickery'.[27]

The question of *māyā* is of importance in any assessment of *ṛta* in so far as this power is used to fit in with *ṛta* or to obstruct the layout of the cosmos. It is one of those words that express the ambivalence which the ṛṣis saw in the universe. It seems to be the meeting point of two currents of ideas,[28] referring to one and the same power which can be used for constructive or destructive aims, for beneficent or maleficent purposes, for forwarding the ends of evolution or impeding them, for working out 'truth' or 'deceit'; hence again the factor of opposition between order and anarchy, creation and dissolution exemplified in this word, each opposing force having its own part to play in accordance with the overruling course, *ṛta*: the difference between the two lying in their purpose or intent. But, in the end, even if *ṛta* is opposed in some way, all will be in accordance with the *Ṛta*, for at the transcendental level *ṛta* can never be thwarted, thouggh it can be so at the phenomenal level.

Some may see here a contradiction in the thought of the ṛṣis, as *ṛta* stands for the world 'order', and the gods' use of *māyā* denotes transformation of that order. This is not so, as the world order is not static but is a constant 'becoming' in accordance with strict rules such as the rhythm to which all are subjected in their manifestation – birth, growth, death, decay – this rhythm itself exemplifying transformation (*pariṇama*, evolution). The world order is both static and dynamic, static in the sense of established – nothing can alter the order of life, the eternal recurrence of day, night, seasons, years, etc. – dynamic in the sense of that constant transformation in accordance with the law inherent in being.

Māyā, in the Ṛgveda, refers to the asuras' creative power, as well as to that power of transformation – later to be used by the demons for deceptive purposes – that power which is subservient to the *ṛta*, or law of transformation. Creation, in the Ṛgveda, is a deed of mighty *māyā*.[29] The world order is in itself magic, it is the asuras' magic for it cannot be fully explained in wholly rational terms.

Agni bears many names and *māyās*,[30] or powers of transformation. Indeed, is he not hidden in the wood, in the stone, in the cloud or water, does he not appear in various guises, as the flame or lightning or solar ray, does he not burn, dessicate and destroy as well as warm, cook, ripen, hence does he not act both constructively and destructively? The beautiful hymn to Earth in the Atharvaveda shows Agni as the substratum of all things:

> There lies the fire within the earth and in plants; and waters carry it;
> the fire is in stone; there is a fire deep within men,
> a fire in the kine, and a fire in horses;
> The same fire that burns in the heavens, the mid-air belongs to this Fire Divine.
> Men kindle this fire that bears the oblation and loves the melted butter.[31]

Earth and Heaven are both *ṛtāyinī*, followers of the *ṛta*, and *mayinī*, i.e. skilled in wondrous art, constantly producing their wonder in changing forms, the growth of plants, the change of leaf to bud to flower back to seed, the metamorphosis of grub to butterfly, of embryo to creature, all these are effects of the *māyā* of Earth's creative work.

By his *māyā* Varuṇa

> has encompassed the night and established the morns (VIII.41.3ab)

thus laying down the law of cycles. The sun which Mitra-Varuṇa 'set in heaven as a refulgent chariot' is compared to a

> flying bird anointed with the asura's *māyā* (X.177.1).

By that *māyā* it has been given the power of circling round the heavens and thereby unerringly follows its course 'not infringing the common law', 'moving forth as light'. By the asura's *māyā,* Mitra-Varuṇa pour forth the rain from heaven. None can hinder the 'mighty *māyā*' of the most wise God by means of which the rivers flow without overflooding the oceans. The laws of nature are viewed as the magic of God.

The very spinning of the web of *ṛta*, reproduced in the Vedic ritual, is attributed to Varuṇa's *māyā*. All thus works with the law which in man's eyes manifests as a wondrous piece of magic. The power of *māyā* itself is the means whereby the world order is constantly expressed and its source is essentially divine, laid down in heaven:

> Your *māyā*, O Mitra-Varuṇa, is laid down in heaven (V.63.4).

Where do the stars that are visible on high at night go when they disappear from the sky by day? asks one poet (I.24.10). The answer that Varuṇa's decrees stand inviolable implies that all works in accordance with his ordinances, i.e. the law. That great power is qualified as splendid and compared to the brilliant moon:

> Mighty is the power of Mitra-Varuṇa, even like the moon it bestows its splendour far and wide (III.61.7cd).

The beauty of the universe and its wondrous workings are magic to the eyes of the beholder.

But *māyā* is not only used to bring out the harmonious movement of natural phenomena. Nor was it only the possession of asuras, godly or demoniac. The Aśvins are also called *māyāvin*. Their chariot is full of magic, or wonder-working *purumāyā*. Their frequent assistance to mankind is a piece of magic. The Ṛbhus and others also resort to this *māyā*. An interesting fact is that the Ṛbhus won their share of the sacrifice and rank among the gods through their skilful artistry and *māyā*. They are described as *māyinaḥ*. Certain mythical prowesses of theirs, miraculous deeds which they performed by means of

1 power of speech (*śacībhiḥ*),
2 vision (*dhiyā*), and
3 mental ability (*manasā* Ṛgv. III.60.2) –

the power of sound, vision and thought capable of producing concrete phenomena – these are summed up in the word *māyā* in Ṛgv. III.60.1cd and show *māyā* as that mental power capable of producing or shaping things at the phenomenal level. Those who are thus gifted are also called *ahimāyāḥ* where the term *ahi*, serpent, may refer to both the wisdom and the wile which the ancients ascribed to the serpent. Thus the Maruts are *ahimāyāḥ* as they 'robe themselves for our welfare in the height of heaven' (X.63.4) as clouds that pour the beneficent rain. Through *māyā* they 'invest themselves' with clouds, i.e. assume the shape of clouds.

On the destructive side, we find *māyā*, the power of transformation, used for deceptive purposes, hence its evil side, though one should not attribute evil to a neutral power, but only to its wrong application, the truth as to good or evil being in the right or wrong motive in its usage. Demons can halt or alter the course of natural phenomena such as the holding back of the waters or of the sun[32] and can change their appearance at will, a kind of glamour plus guile or craftiness and stratagem calculated to throw their adversaries into confusion but which the gods, with the same power at their disposal and by also donning various forms, can easily counteract.[33] This aspect of *māyā* inevitably led to the meaning of deceit, thence illusion, to the exclusion of all other meanings and is found far more developed in post-Vedic literature than in the Ṛgveda. In the final analysis, however great the power of the *mayināḥ* or of the *dhīrāḥ*, (wise), that power cannot go against the statutes of the gods who are the embodiments of the *ṛta*.

> Neither those who possess *māyā* nor those who hold wisdom can impair the primeval, firmly-fixed statutes of the gods.[34]

So by means of *māyā* the world order seems capable of unsettlement and by means of that same *māyā* that order is re-established, for in its deepest aspect nothing can alter that which is to be established by *ṛta* and is already so established at the transcendental level. Thus is the universe wrought, according to the Vedic vision, and kept going, the root of this constant organising and disorganising being found in the original interplay of two opposing forces which are necessary to bring the universe into existence. That it is a cosmos, or field of ordered activity, is due to the law of its very being that acts as soon as the first impulse to manifestation has been given, and keeps all opposing forces in equilibrium. That the forces of chaos keep threatening the cosmos is also part of that vision that finds its proof in nature: all the glorious constructions of man to his gods, when left to themselves, are soon buried by jungle growths or sands and finally fall into decay. Mind, or man and deva, is that which constantly fights to keep the order.

The divine encounter

The vision of the godly realm is on the whole one of splendour. Like

> Dawn, the shining daughter of Heaven . . . dispelling gloom of night, ushering in the light (VII.81.1)

so the ṛṣis in their contemplation lifted the veil upon a luminous world seldom envisaged by the human mind. Above all and in spite of certain disruptive forces and one or two discordant notes,[35] it is a vision of cosmic harmony and solidarity, the wonder of the working in unison of all the great energising powers of the universe as they shape the ideal thought of the Creator, the Supreme Ordainer (*dhātā*). These cause the universe to manifest according to certain patterns of progressive unfoldment and harmonious expression in which every realm of activity, supervised by a god, develops in perfect concordance with every other realm. The universe, at the godly level, is thus the dynamic expression of *ṛta,* all the gods' action moving concertedly towards the one end of creating a harmonious world, and this is reflected at the phenomenal level in the harmony of nature:

> All gods, one-minded, one-intentioned, unerringly proceed to the one purposeful accomplishment.[36]

The infinite manifestations of nature all converge, in the poet's heart, into one mighty paean of beauty. Harmony is beauty, beauty is harmony:

> In the stimulation of Savitṛ let us contemplate all beautiful things (V.82.6b)

sings one poet. Another's address to Indra is intended to give a glimpse of the richness of the god of whom nature is the reflection:

> Thou who . . . hast made . . . the earth the counterpart of thy creative might.[37]

Free from all human failings, pure, infallible, undeceivable, the gods of the Ṛgveda represent truth (*satya*), they are the 'lords of truth', 'faithful to the law', the recipients of the vision of *ṛta* and its very embodiments; theirs being the dominion of light, wide space, freedom, of glory, brilliance, power, of benevolence and splendour, all of which sum up the wonder and admiration in which the Vedic bards held them. The gods' action is constantly described as 'of one accord' (*sajoṣasa*) or harmonious; 'acting in harmony', in 'unanimous accord', the gods have awarded their dominion to Mitra-Varuṇa. Similarly, there came a time when to Indra's 'divine dominion the earlier gods granted their powers'.[38]

Accordant and of one mind (*samanasaḥ*) they all come to the worshipper to bless the sacrifice. The quarrellings, jealousies and hostility that characterise the gods of the later Hindu pantheon are hardly incipient in the Ṛgvedic pantheon.[39] Here the gods are and remain the

> celestial denizens of heaven, flame-tongued, thriving through the law, abiding brooding in the womb of law.[40]

No wonder that the bard could exclaim:

> Blessed is all that the devas regard with favour.[41]

The whole hymn to earth in the Atharvaveda (XII.1) is a panegyric of harmony on earth. As already noted, the universal harmony, the gods' handiwork, is their song in which they rejoice:

> The gods revel in the sacred-song (*sāman*) of *ṛta*.[42]

This song may be worded by the bards but who inspires it?

The cosmic order is itself the divine song of the gods which the poet intuits and expresses in his own words at the sacred function of the sacrificial ritual.

The essential oneness of manifestation as rooted in the divine harmony, *ṛta,* is expressed in far-sweeping imagery taking in the whole of nature, as the divine swan with its abode everywhere, the swan which is *ṛta*:

> Swan, dweller in the light, dweller in the inner space, beneficent priest present at the altar, guest in every dwelling, dweller amongst men, dweller in the wide-spaces, dweller in *ṛta,* dweller in heaven, born of waters, born of kine, born of *ṛta,* born of mountain, (such is) *ṛta.*[43]

The gods are regarded as gracious towards humanity, for they are of 'one spirit' (*samanyu*) with man and are 'bounteous to every man'. Their further task is to train man to reflect this harmony, their own peculiar achievement, in his own domain. It is here that we meet man and his collaboration with or disruption of the divine solidarity, the 'pattern in the heaven', *ṛta.*

Part III

THE HUMAN LEVEL: MAN'S PLACE IN THE UNIVERSE

INTRODUCTION

At the third or most concrete level of manifestation, we encounter man. Just as *ṛta*, in its universal connotation, is the law that brings about order out of chaos and creates harmony, so in its ethical aspect it governs man in a similar way. To follow the norms of *ṛta* is to create order in human relationships. That order which mirrors the cosmic order is truth. Truth, integrity, righteousness, these are the human expression of universal harmony.

The idea of universal law becomes, at the human level, particularised into specific laws. Thus are developed those norms of social order which are meant to bring the best results for all concerned. Therefrom is born civilisation. So one of the most important aspects of *ṛta*, in so far as man is concerned, is its ethico-social side. Here can be traced the root of the doctrine of *dharma* – duty or moral obligation, of one's function in life or vocation, and that of *karma*, or the doctrine of action and reaction as it affects man, the effect of causes brought about by human action. The conception of *ṛta* as the right course of things mirroring the supreme harmony inevitably led to these two specifically Indian notions of *dharma* and *karma*.

Man in the Vedic vision is part of that cosmic whole, a dynamic part, a disturbing factor, yet one called upon to play a vital role in the ultimate establishment of the law in the manifested world. He claims nothing short of integration, glory, splendour:

Undivided am I, undivided my soul,
undivided my sight, undivided my hearing;
undivided my in-breathing, undivided my out-breathing,

> undivided my diffusive breath; undivided the whole of me (Athv. XIX.51.1).
>
> As glory let my oblation increase quickened by Indra, of thousand-fold might, well brought, made with power . . .
> Unto our glorious Indra, rich in glory, would we, rendering homage, with glories pay worship; do thou bestow on us royalty quickened by Indra; in the bestowal here may we be glorious.
> Glorious was Indra, glorious was Agni, glorious was Soma born; glorious of all existence am I most glorious (Athv. VI.39.1).
>
> On this splendour have all the powers of heaven concentrated their thought. May those gods who nourish all life on earth anoint me with splendour (Athv. III.22.2).

These are only samples of that yearning for glory, splendour, scattered in the collections, a glory that is not the fruit of warring but that of the illumination of the spirit, perhaps of that martial spirit that allows man to conquer all obstacles till he reaches his supreme goal. As Dr Panikkar rightly remarks in *The Vedic Experience*:

> Human glory seems here to be less an imitation of the divine glory than a participation in the universal and indeed divine splendour of creation. Man's glory is not a reflection of God's glory but a sharing in the very splendour of God by a full realisation of the human function in the universal adventure of reality (p. 338).

Integration is the goal of man for he is that being that unites within himself all the opposites. A quaint passage in the Atharvaveda, quoted below, brings this home to us. Obviously the wholeness of the cosmos can only be mirrored in integrated man, man made whole, at peace with himself and the universe, whose faculties and powers are gathered together in one central point wherefrom he acts in full consciousness, unperturbed and balanced. That is man's goal. But before that can be attained there stretches before man a long path of striving and erring, of suffering and aspiring, of frustration and achievement, of falling back and rising again, of endless endeavour which indeed forms the pattern of his glory. Not mere achievement is glory, but every step and every surmounted set-back adds colour, depth, weight, to the end result.

In one of the most important hymns of the Atharvaveda (XI.8), we find a description of man as the abode of godly powers, man as the child of the deities, man as the seat of the opposites. This in itself sums up the glory and the predicament of man. The gods fashioned man; then,

> Having poured together the whole mortal the gods entered man (Athv. XI.8.13cd).

All the parts of the human body enumerated in two stanzas are said to have been assembled together, and the question is asked

> What seer put that together? (14)
> Who brought into it (the great body) the colour with which it shines here today? (16)

Then sleep, weariness, dissolution, deities named evil, old age and so on enter the body, as well as all the ills to which man is heir, and the opposites quaintly described as

> Both growth and diminution, generosities and niggardlinesses, both hungerings and all thirstings (21)
> Both revilings and non revilings (22) both knowledges and ignorances (23)

and then the *brahman* entered the body (23)

> Both blessings and precepts, demands and explanations, thoughts, all divisings, entered the body afterwards (27).

Stanza 30 repeats that

> brahman entered the body and on the body is Prajāpati.
> Therefore indeed, one who knows man (Puruṣa) thinks

> 'this is brahman'; for all deities are seated in him as cows in a cow-stall (32).

Thus equipped, is man not the child of the cosmos, the microcosm of the macrocosm, capable of mirroring in his life, as well as in his thought, conception and vision, the polarities that make the web of the universe, as well as of integrating them in himself through the power of the *brahman*, the source and goal of all things? The vision of the ṛṣis bears witness to this claim.

SIX

The meaning of *anṛta* (disorder, chaos, falsehood) and the role of dissolution (*nirṛti*)

In the conception of *ṛta* and its opposite *anṛta*, the dual aspects of metaphysical and ethical, and of positive and negative evaluations have to be considered.

Ṛta signifying order, light, form, right, *anṛta* would point to the opposite, disorder, darkness, formlessness, wrong, each word thus bearing two strands of connotations, metaphysical when applied to the cosmos and ethical when applied to man. Both metaphysical and ethical notions are closely intertwined as will become clear in our examination of *anṛta* meaning disorder, darkness.

The sun is said to uphold *ṛta* and dispel (or overpower) *anṛta* (I.152.3). With his appearance he puts an end to darkness which represents the formless and unorganised or undifferentiated world; he ushers in the organised universe. Similarly, when spiritual light shines in human consciousness it puts an end to all sorts of wrong attitudes, in short, to human darkness, as also to the undifferentiation of vague feelings and thoughts. As pointed out by Louis Renou, the sun is the support (*skambha*) of the world, 'whose presence signifies precisely the ordered cosmos, heaven and earth duly separated, measured out.'[1]

But we should beware not to ascribe to the word darkness, when used in a metaphysical sense, any ethical negative connotation such as evil or wrong. Darkness is repeatedly said to contain light; it is that unknown (to man) container, the matrix wherein light, form and order have their birth. On the other hand, darkness and light seem to be in the same relationship as the two opposite poles, spirit and matter, which, when fused, constitute the unmanifest (*asat*) and which, when duly separated, make manifestation possible. It is the task of the gods to separate their intermingling, to unravel them, thereby making cosmos out of chaos. A number of verses bears on this idea:

> Brahmaṇaspati, seeking light in the darkness, drove out the morning-beams as cows standing concealed in the bond of anṛta (X.67.4bcd).

Brahmaṇaspati, the lord of prayer, plunges into the darkness to find the light. Darkness and *anṛta* are here identical; out of chaos or darkness, the morning beams are extracted, the herds of the universe emerge, take shape, become perceptible as the light grows. A similar idea, expressed by mythological language, is found in the myth of the mountain Vala, cleft by means of prayer (*brahmaṇā*) whereby the darkness is dispelled and heaven's light revealed (II.24.3). Again with his rays of light, the Lord of prayer forces apart the darkness (X.68.9) and

> mounts the refulgent chariot of ṛta (II.23.3b)

chasing away the darkness, opening the cowpen and finding the light of heaven (*svarvidam*). Indra, on his part, by means of the light of his thunderbolt, milked the cows out of the darkness (I.33.10) and again

> rolling down the darkness by means of his light, he drove forward the milky cows from within the cave (V.31.3cd).

He finds the sun lying hidden in obscurity, he too discerns the light from the darkness.[2]

These examples, couched in mythological language, tell a tale of light taking birth from the womb of darkness, the cosmos arising from the chaos. The conflict between light and darkness is thus the struggle of light out of the womb of darkness which encompasses and holds it hidden, as of the spiritual out of the material depths in which it is imprisoned. The forces of darkness are considered 'obstructing' and those who embody these, such as the Vṛtras, the Paṇis, come to have an evil connotation; they are obstructors, but in the long run cannot win, for what has been decreed as the prevailing order must conquer.

Mythological language has its own peculiar terminology and uses personifications and concrete symbols, of which the word 'cow' (*go*) is an example. This word has various meanings quite different from its literal sense.[3] Aditi, the goddess of the Infinite, and Dawn, the goddess that ushers in the lord of light, are called 'cows'; Dawn is the 'mother of the cows' (IV.52.2); the luminaries that speed across the sky are cows or herds, and it has also been suggested that the disputes between seers and nobles concerning 'cows' may refer not to actual cows, but to women, 'cow' being a complimentary term in India, to this day.

The forces of darkness cannot have meant sheer evil for we find the curious but quite revealing reference in the Atharvaveda to the sun as 'born from the ocean, born from Vṛtra'.[4] The waters of space, the subtle essence of matter out of which the universe is fashioned, are the darkness, matter and womb of the universe, here identified with Vṛtra, the dragon of matter. In what man considers darkness or chaos, because it is beyond his comprehension, lies hidden the principle of light, of order, of cosmos, of life-giving energy:

> In the waters are immortality, in the waters are healing balm (I.23.19).
> In the billowy deeps [is] concealed the sun (X.72.7).

The theme of a hidden treasure that grants life and light is a recurring one in the Ṛgveda, as indeed in all religious scriptures, whether the treasure be the sun or cows hidden in caves of mountains or the waters of life released as a result of the slaying of the dragon Vṛtra that held them back. The 'treasure of heaven' which Indra found is said to be

> fast hidden as the bird's young in the cave encompassed by never ending rock (I.130.3ab).

Similarly the 'highest treasure' is hidden and secretly hoarded by the Paṇis and the efforts of the patriarchs to find it make the subject of a myth.[5] Agni, also, is said to have

> cleared the doors of the Paṇis for us and revealed the life-nourishing sunlight (VII.9.2ab)

that 'bright wealth that shines brilliant and powerful among men' (II.23.15) and that is prayed for to the Lord of Prayer, the son of *Ṛta*. Louis Renou interprets such withholding of treasure in terms of engendering disorder: 'It is disorder which the Paṇis engender by keeping the cows (symbol of light II.24.6 and 7) prisoner. Such an act disrupts the regular march of the cosmos.'[6] To 'withhold' as against 'sharing' and 'giving' was considered a grievous sin. The universal order is bounteousness. Nature is constantly giving of her bounty to man. Thus generosity is one of the highest qualities and any behaviour that runs counter to it is 'sinful'.

A psychological meaning may be found in this mythological language of bursting mountains open, releasing cows or waters and setting the sun on high; namely the longing to cleave through the mental darkness or limitation which prevents man from seeing further, from seeing the light and penetrating into the beyond; man is himself the rock that contains the treasured light. This may be hinted at in the following verse:

> May we be sons of heaven . . . and shining forth break open the treasure-possessing rock.[7]

It is not enough that the patriarchs should have performed these prowesses in the past: their descendants wish to repeat the feat here and now.

Darkness, in its metaphysical sense, is not evil and could not have been considered as such in the early days of Ṛgvedic civilisation; it is both the complement of light and that which contains the light which is the life of man. For this, Vedic man could think of no better symbol than the cowpen 'in the bonds of *anṛta*' waiting to be opened up, or of the sun hidden in the rocky cave waiting to be released and set up in the sky. *Anṛta*, in such mythological tales, probably means chaos. But darkness's reflection, when ascribed to the human order, becomes the symbol of evil, of constriction; hence also Vṛtra, Vala, the Paṇis, those that hold back, constrict or obstruct, those that hide and hoard the treasure, hence the development of meaning of *anṛta* from 'disorder', 'chaos' to 'wrong', to 'false', to 'untrue', which last it kept throughout the classical period.[8]

Anṛta acquired the meaning of falsehood, untruth, wrong as a secondary meaning to disorder, to *anṛta* as opposed to *ṛta*, order, possibly early in Vedic times. That which is not in accordance with the divine order of things, as seen by human eyes, is wrong, false, and conduces to suffering through the friction created thereby, hence to evil. Those who are *anṛta* and *asatya*, i.e. false and untrue, are 'sinful', whilst those who are

> observant of the *ṛta* are truthful (VI.50.2c)

since they are in harmony with that which is the expression of truth. When coupled with *satya*, truth, *anṛta* means only falsehood and was to be kept as the opposite of truth (*satya*) even when the word *ṛta* itself fell into disuse.

At times there seems to be confusion in the poet's mind as to what is the right and what is the wrong as established by the gods' ordinances, or where to draw the line between what belongs to the cosmic order, hence the truth, and what to the disorder brought about by human infringement of divine law, hence the false. He asks:

> Which is your *ṛta*, which your *anṛta* (I.105.5cd).

The previous verse seems to hint at a time of change for the poet wonders:

> Where has the ancient law departed? Who now upholds it?

To what kind of *ṛta* is he referring? The social order, ethical norms, or ritualistic practice which may be disregarded? The separation between *ṛta* and *anṛta* is not always clear, both seem to be closely interwoven just as the sun, the cows and dawn are said to be hidden in the rocky cave. What is considered right for one social order loses its value in the next and this, in transition times, causes undue bewilderment.

Can the universal order be impaired in any way? Such a verse as the following might be thought to imply it:

> Let thy noose, O Agni, thrice ensnare the evil-spirit that by *anṛta* strikes at *ṛta* (X.87.11ab).

But this invocation may simply refer to the sacrificial rite which is being thwarted by an evil spirit. *Anṛta* would therefore imply contempt of the ritual, a contempt which, in its wider connotation, would denote scorn for the values of orderliness, harmony, worship and respect for the divine laws. It may also imply that *ṛta*, in its particular applications, such as the norms of the social order or the ritual, may be attacked by those who refuse to follow these norms, by those separative, divisive forces that work against the universal order in contradistinction to the constructive, harmonious forces that work for it. But since *ṛta*, at the transcendental level, is itself the ultimate harmony behind both currents of forces, the unitive and the divisive, in the final analysis nothing can prevail against it, though at the phenomenal level conflict must go on from the very fact that the universe is structured on the tension of two opposite polarities. In particular and practical applications, wherever disturbances do occur, these must be rectified in accordance with the law. The constant bringing back to equilibrium, to harmony, is part of the course of things and in the moral sphere, where humanity is concerned, it heralds the doctrine of karma, as will be discussed in the section on karma.

So Agni's fetters await 'those who protect the seat of anṛta',[9] for these in favouring, or serving, the wrong, bind themselves to it and thereby lose their true freedom.

Dr Panikkar's assessment of the Vedic notion of evil sums up the whole position:

> It is a distinctive feature of the Vedas that responsibility for the existence of human suffering is not transferred to a Prince of Darkness in one form or another; rather, evil is taken to originate in a malfunctioning of the given structures of reality, owing to a clash or conflict of interests. Evil is situational, we may say, and not ontological.[10]

That hell, destruction, dissolution, have their place in the Vedic vision of the Cosmos is evidenced by some references in the hymns, especially in imprecations against demons, but this aspect of manifestation is not insisted upon; indeed the exact meaning of these terms is far from clear. Such expressions as 'bottomless darkness', 'deep pit', 'chasm' to which demons are sent, or the 'lap of destruction' into which the forlorn lover would sink, or which, as the personification of the goddess Nirṛti to whom homage is paid in the Atharvaveda, is not always considered nefarious, but is again and again requested to keep away from the worshipper in the far distance – all these are capable of different interpretations and do not lead to any clear cut conclusion.

As L. Renou says, 'The Vedic hell is first of all disorder, that which breaks the regular course of things and destroys the harmony.'[11] It is not hell in the Christian sense but its 'equivalent', with the vast difference that there is no evidence of a place of eternal torture, or of annihilation, as some scholars believe.[12] Let the demon or sorcerer who impedes the righteous work of the priest go to non-existence, or to destruction, or fall into the endless pit or bottomless darkness, into the abyss below all creation; let your furious rage, O Indra-Soma, conquer him; these are samples of the imprecations hurled against the hater of the *ṛta*, he who works 'with charms counter to the *ṛta*' (VII.104).

The question of the meaning of *nirṛti*,[13] dissolution, and its place in the Ṛgvedic cosmos is still moot. The word itself has been taken in opposition to *ṛta* but its connotation is different from that of *anṛta*, the latter meaning disorganisation in the sense of disorder, the former, i.e. *nirṛti*, meaning disorganisation in the sense of dissolution and therefore destruction. For L. Renou the two pairs of opposites, *ṛta* and *anṛta* belong to the ethical plane, and *ṛta* and *nirṛti* to the cosmological plane:

> There is an opposition between ṛta and anṛta, disorder or falsehood, on the ethical plane, and between ṛta and nirṛti, dissolution on the cosmological plane. The power of the gods is limited by the interplay of these forces, just as later it is limited by karman or māyā.

These cosmic intelligences, being embodiments of powers that conduce to order or dissolution, are therefore the dynamic expressions of two forces, of construction and destruction, cosmos and chaos. This is where lies the root of the problem of the asuras and their so-called beneficent or maleficent action, their unitive or separative influence on the cosmos. *Nirṛti* has its role to play in the vast scheme of becoming; within certain bounds it also forms part of the ordained course of existence since transformation involves both building and dissolving of outworn forms. Gods – as opposed to demons – and men do not stand on this destructive side of the cosmic process, although they may have to resort to destruction to serve the ends of construction. Thus Indra's slaying of the dragon Vṛtra, emblematic of static nature, results in the release of the waters of life, the generation of the dawns, the setting on high of the sun, the shaping of darkness into a cosmos of light and, in so far as man is concerned, the granting of freedom. The rendering powerless of the forces of inertia or obstruction served the higher purpose of the forces of evolution. But the interaction between these two forces is necessary that the universe may be. Hence destruction is part of the cosmic scheme, though dreaded in a sense by man whose very constitution is a compound of many elements and for whom destruction or dissolution obviously means the end of his person as such, but who nevertheless plays his part in bringing about destruction in the universe. However, since that is not apparently his *dharma*, it can only result in suffering and evil to his kith and kin. Destruction and death are usually considered calamities from the human point of view and are dreaded.

Nirṛti does not seem to be completely identified with death in the Ṛgveda, though there is ambivalence in some passages and in post Ṛgvedic literature it is so identified. It is rather the process of disintegration that sets in after the severence of life and body has occurred. It seems therefore to signify and indeed personify the forces of disintegration that go against the ordered 'layout' or organisation (*ṛta*) of the universe, the process of decomposition that sets in when life is withdrawn from a body, indeed the forces of putrefaction both physical and moral, the complete disintegration of the body or the person, those forces of disorder that are in constant conflict with those of construction and order, both in the universe and in the human constitution, forces which, in the latter case, take over completely and cause the annihilation of the physical frame as such after death. *Nirṛti* thus represents the Ṛgvedic conception of that tendency, prevalent in nature, towards disorder and disintegration – a direct effect of the discontinuous structure of matter – which modern physics translates in terms of the laws of degradation and measures through what it calls entropy. *Nirṛti* could then be a personification of what we would now call the law of degradation. Such forces are also prevalent in the moral and psychological sphere. The following verse can be read from both the physical and the moral point of view:

> Let not baneful *nirṛti* rule over us (X.36.26).

There is no direct evidence that the 'abysmal station' which those who are full of sin have engendered for themselves (IV.5.5) is equivalent to *nirṛti*, but the latter is, in the Atharvaveda, classified together with certain categories of beings and states of consciousness which are far better avoided:

> From the niggard, from perdition, from sorcery, also from fear, from the violent deadly weapon of death, let (the magic formulae) shield thee (Athv. X.3.7).

Men labour under the fetters of death[14] which are used as curses against enemies; so *nirṛti* is invited to bind foes with those 'unreleasable fetters of death'[15] but sometimes the knot which *nirṛti* ties around the neck of man can be released through – as it seems – the priest's power of invocation:

> Homage be to thee O Nirṛti, thou of keen keenness; unfasten the bond, the fetters of iron. Yama verily giveth thee back to me! To that Yama, to death, be homage.[16]

The wish is also entertained that earth preserve the dead man from *nirṛti*'s lap – probably an allusion to the decomposition of the body (X.18.10).

The whole idea of the forces of construction or evolution working *pari passu* with the forces of dissolution is found in two somewhat enigmatic stanzas which are capable of various levels of interpretation:

> Whoever has made it does not know it;
> whoever has seen it surely is hidden from it.
> He encompassed within his mother's womb, source
> of much progeny, has entered into dissolution (I.164.32).[17]

At the cosmological level, the mother's womb may refer to Aditi in whose lap hides the seed of the universe, source of progeny, this word signifying activity, and activity birth, growth, dissolution; to enter into activity is to enter into the realm of dissolution, *nirṛti*, to create and to abolish. This is indeed part of the eternal rhythm, hence *ṛta*. At the human level, the one who is the cause of the birth of the child does not know the secret behind this birth, even the external signs cannot help unravel the ultimate secret of the momentum and the *modus operandi*. The seed is source of progeny but must itself die as such in order to become that which it contains in essence, hence enters the realm of dissolution in order to give birth to itself. As hinted in Ṛgv. X.114.2, the next enigmatic stanza to be examined, these forces of destruction which are at the root of the universe can themselves teach the sages the secret of life and death:

> Those oblation-bearers who discern them, as they are heard from time immemorial, serve the three *nirṛtis* for instruction. The sages have perceived the connecting link

> of these within the secret and remote inroads (of the universe) (X.114.2).

The forces of dissolution are here admitted to be inherent in the very structure of the universe which is the field of compounded elements and therefore of construction as well as destruction, both being aspects of the course of things, or order, *ṛta;* though from man's point of view, construction and order stand for *ṛta,* the right, whereas destruction and disorder stand for *anṛta,* the wrong, it is obvious that metaphysically, the problem is far more complex. From a wider standpoint, the forces of organisation and of disorganisation equally express the established order as a balance between two opposite tendencies, the balance and its keeping being *ṛta.* Birth, growth, death, decay, the breaking up of forms into their component elements, the disintegration which affects every structure, all proceed in accordance with law; this is the vast becoming of the universe, the eternal transformation that characterises manifestation, or the universe to which the sages referred as 'this all'.

The above quoted verse seems to imply that the sages hear the ever-repeated sound emitted by those forces of destruction and discover their *raison d'être* in the very law of manifestation to which they are subservient. Hence the priests serve these *nirṛtis* for the sake of receiving enlightenment into the secrets of the universe, into the forces of involution and evolution. L. Renou explains the meaning of the verse thus:

> These three nirṛtis are those factors of entropy . . . the forces of rupture of the ṛta on the various levels where Vedic speculation moves; celestial, aerial, terrestrial space . . . divine, ritual, human activity – a de-structuring power, the polar counterpart of the great positive entities of Vedic thought.[18]

All around us, *nirṛti* prevails, for all is subject to dissolution, yet life goes on, Indra reigns supreme (VII.37.7a). A further example of this 'destructuring power' that has its deep *raison d'être* in the universal economy may be seen in a description of the Maruts as

> pounding the two worlds with their mightiness as from the point of dissolution and the pillarless-region they reach out to the firmament (VII.58.1cd).

The Maruts take in their stride the world of dissolution, or terrestrial realm, and the firmament or the mid-region as well as the heavens. But is there here another meaning to *nirṛti*? Is the point of dissolution the critical stage of transformation from one state of being to another, (*laya* centre), from one realm (*loka*) of being to another?[19] From that point of dissolution they reappear to another level, thereby contributing to the constant transformations to which the universe is subjected. We can only surmise.

The Aśvins are said to have delivered a certain man (Vandana) from the pit like a hidden treasure, to have brought him forth like buried gold,[20]

> like one who slumbered in destruction's bosom, or like the sun when dwelling in the darkness (I.117.5cd).

If the darkness in which sleeps the sun is taken on a par with *nirṛti* and if both are similar to the cave whence one extracts treasures, then *nirṛti* for all its dissolution process is the state and place of transformation, where the forces of nature having disintegrated to their maximum are slowly being re-integrated to mould another form.

From decomposition emerges re-composition, from dissolution new life; through that power that is both dissolving and re-solving the world is constantly being transformed, born anew; the seed grows through a process of annihilation of itself as seed through the dark soil, the embryo in the dark womb. Likewise grows man, from darkness to more and more light, from ignorance to wider and deeper knowledge, from vice to virtue, from brute to sage. Even sin is an experience out of the darkness and heaviness of which he emerges victorious to claim his birthright, freedom, knowledge, wisdom. The cosmic order is the dynamics of becoming and man its epitome.

SEVEN

The concept of sin in the Vedas

Sin, which is a purely human characteristic, results from the violating of *ṛta* in the moral sphere. Through weakness or ignorance, through his 'human estate'[1] man violates the great Law. Why is that? It was recognised that man belongs both to the highest and the lowest, that he has within himself both the divine and the brute, the lofty aspirations, inspirations and self-oblivious actions, and the most deadly, calculated, egoistic deeds. The one aspect grants him that faculty which enables him to realise what is right, the other, in its separative selfish drive, pushes him to commit what from the higher point of view is wrong, or to ignore the right which is set before him; hence his lower nature tending to corruption – his weakness – and his higher nature, holding *in potentia* his spiritual strength. In Ṛgvedic terms, he is that mortal in whom dwells as 'guest' Agni, the flame divine who

> raises the mortal to highest immortality (I.31.7b).

It is this that makes him a link between the terrestrial and the spiritual.

In his corruptible state man often fails to grasp or discern between right and wrong, or finds all sorts of excuses for his wrong doings, excuses which afford us glimpses into Vedic society and beliefs, such as that one's sins are also the sins of ancestors, so that the sins of the fathers are visited upon the children; the causes of, or excuses for, sin being delusion, wine, anger, dice or thoughtlessness.[2]

But in the human heart is placed that deeper understanding which characterises human intelligence, (*kratu*), and which is a gift of Varuṇa.[3] Man thus has the capacity to know what is right and what is wrong although he may be so deluded as not to be able to discriminate between the two, or may turn his back on the right. Varuṇa 'opens out in the heart an inspiration' and thereby 'brings about a new (vision of) *ṛta*'.[4] So man knows, in his heart, the meaning of transgression which is essentially a disruption of the established harmony. He recognises the eternal law of the gods and proclaims it.[5] Those who observe the *ṛta* follow the ordinances of the gods as established by the divine laws and thereby enjoy the protection of these; and it is through the upholding of religious ordinances that the patriarchs became illumined and inspirers of their descendants.[6]

Man's transgression is a debt[7] (*ṛṇa*) which he contracts at the hands of the guardians of *ṛta*, the Adityas – whatever moral transgressions men commit are considered sins against the gods – a debt which man must pay in order to re-establish the balance he has upset. The Adityas, 'true to *ṛta*', are the 'debt-exactors'.[8] Indra is he who demands fulfilment of one's obligations. Man pays his debt to the gods as agents of the great Law, should he infringe its decrees, by suffering the effects of causes set up through his wrong doing, whether this be committed in knowledge or in ignorance. This obviously is an aspect of the law of cause and effect which is not explicit but always implicit in the Ṛgveda. In the process he may accuse the gods, in his ignorance or anger, but he knows that fundamentally those who infringe the statutes of Mitra and Varuṇa for example

> injure as it were a friend (X.89.8c).

The gods differ ethically from man because they never infringe the one Law of whose statutes they are the guardians, whereas man does. But man often claims his common origin with the gods and seldom cringes to them.

As all works towards the greater harmony and therefore well-being of all, the ultimate criterion of *ṛta*, whether at the natural, human, or godly level, as ever, is harmony. The moral imperative is the expression, at the human level, of the supreme harmony: all human relationships that conduce to harmony, unity, are expressions of truth, all those that lead to separation, division, lead to suffering and evil.

The moral significance of *ṛta* is usually considered to be centred mainly around Varuṇa and man's relationship to the great God of Cosmic Order; but it can also be found as dynamic in its relation to the other Adityas, and in addition to Agni, to Soma, Rudra, Indra, Heaven and Earth and Dawn, the Waters, Aditi, Savitṛ, Sūrya, Brahmaṇaspati, all of these being the guardians of the law in one aspect or another and the great helpers of mankind:

> For of one spirit are the gods with mortal man, co-sharers all of gracious gifts (VIII.27.4ab).

Such a statement brings it home to us that in the Vedic idea gods and men are brothers, i.e. born in the 'same mother's womb' (VIII.83.8) of the same father. They are related.

But 'the ultimate moral and religious imperative, the source of the statutes of the Gods', as rightly noted by H. Lefever, is *Ṛta*, 'the cosmic and dynamic principle of Law and Order'.

> Ṛta in the natural world is only a manifestation and never a complete manifestation of the transcendent reality (cf. I.139.2). Ṛta, as objective unity, is transcendent, not only in relation to the world but also in relation to the Gods. Ṛta as the actual order realised in the world, is a product of the activity of gods and men, an activity directed in accordance with the transcendent ṛta.[9]

The Adityas are, as we have seen, intimately connected with its actualisation in the phenomenal world, and in this task man has equal share.

Several terms are used in the Ṛgveda in the sense of transgression against what is right, some meaning 'having gone against', or 'transgressed', hence what is morally owed, a 'debt'.[10] Such a variety might be taken as an indication that the ṛṣis may have given far more thought to man's conduct and transgressions than some scholars are prepared to admit. There is also a great variety of differences in the assessment by scholars of the Vedic attitude to sin, nineteenth century Orientalists emphasising the almost non-existent consciousness of sin which their Christian prejudices would have them see, others, like R. Shamasastry,[11] going to the other extreme. A more sober assessment was given by H. Lefever. According to him, the Vedic bard 'realises that the favour of the God is absolutely conditioned by his own righteousness. To regard adversity as the inevitable consequence of sin is to display, not a weak, but an exceedingly vivid consciousness of the gravity of sin.'[12] Whether adversity is thought of as caused by sin, or not, it can also be regarded as a field of experience in which the human 'metal' is forged. In whatever way it is looked upon, however, the consciousness of sin did not morbidly affect the usual optimistic Vedic attitude to life, its keynote remaining one of vigour of spirit and *joie de vivre*. Worldly prosperity seems to have been viewed as one of the signs of godly favour and adversity of godly displeasure, each being the effect of man's action.[13]

Reward and punishment are human value judgments applied to what essentially is action and reaction. These, when translated into terms of reward and punishment can only be 'meted out' by more or less personal agents of the law, who in this case are the gods.

The Ṛgveda is not a handbook of ethics. The hymns, it should be remembered, are primarily invocations and praises addressed to various deities, not a treatise on morality, and therefore any reference to ethics would be purely incidental and linked to the poet's feelings or state of mind at the time of composition; his main preoccupation being to invoke the presence of the gods or to shine in excellence among other poets. Certain states of mind and feeling reveal anguish at having committed some transgression against the law of which the gods are the custodians, hence, against the gods themselves. From such hints one may glean a fair idea of what was considered right and wrong.

Viewing the subject as a whole, there seem to be two outstanding kinds of sin: to sin against generosity, and to sin against truth. Right and wrong may be reduced to two pairs of opposites: straight and crooked, and true and false.

The Vedic ethical ideal seems to have been a combination of generosity and truthfulness. The man who gives everything, or his very best, cannot be deceitful, he is not prone to withhold, keep apart, but is open, truthful. Conversely, he who speaks the truth is not prone to hide or keep back. Miserliness, withholding, keeping hidden, are regarded as abhorrent as much as lying, cheating and deceiving. Generosity and truth, to the Vedic mind, go hand in hand. It could also be observed that the notion of generosity may have entailed its corollary, complete self-sacrifice even unto death. A gift that withholds nothing is in a certain sense a perfect sacrifice. The great emphasis on sacrifice may also have stemmed from this code of ethics. That it applied most probably more to the ruling than to the priestly class is another question.

The three categories into which Dr Panikkar divides the Vedic assessment of sin and evil are useful for a deeper study of this question. He finds 'three basic insights which, like three seeds, may later develop into three complete theories'. The first series of words[14] that concern sin indicates an 'external cause' for evils of all kinds. The second series (anguish, darkness) denotes an 'internal source': 'These words speak of anxiety, narrowness, lack of expansion, a feeling of imprisonment within oneself. Here the cause is within and seems to be inherent in our own nature.' A third series of words[15]

> seems to suggest that evil springs from maladjustment and malfunctioning of a system that otherwise is far from being bad. These words postulate a kind of factual ambivalence in almost any human values which can turn out to be either negative and mischievous or positive and beneficial.[16]

The relativity of every human value is part of the Vedic realisation of the ambivalence prevalent in the universe.

Lying, cheating, harming, hating, greediness, miserliness are all sins highly reprobated. Incest is referred to as sinful in the famous dialogue between Yama and Yami, the first brother and sister, and as against the *dharma* of Mitra and

Varuṇa.[17] It is also *anṛta*, against the great law. This is enough hint to show a strict moral code in sexual matters. Loyalty to friend is highly prized; he who abandons his friend can no longer hear the divine voice. He who transgresses in any way the path of order ceases to be part of the divine harmony and fails to hear the divine word.[18] Sharing is an expression of friendship; offering the best to deity is man's way of sharing with his god. All men come together in the ritual to share in the offering. If they fail to do so in the proper way, they will fail to invoke the deity, they will cease to hear the word. As a corollary, sharing with the needy is a must:

> The gods have not ordained hunger to be our death; even to the well-fed man comes death in varied shape. The riches of the liberal never waste away, while he who will not give finds none to comfort him.
>
> The man with food in store who, when the needy comes in miserable case begging for bread to eat, hardens his heart against him – even when of old he did him service – finds not one to comfort him.

It is recognised that physical ailment, such as hunger, even if considered the punishment for some sin or other, is no excuse for the refusal of the better off to feed the hungry:

> Let the rich satisfy the poor implorer, and bend his eye upon a longer pathway. Riches come now to one, now to another, and like the wheels of cars are ever rolling.
>
> The foolish man wins food with fruitless labour; that food – I speak the truth – shall be his ruin. He feeds no trusty friend, no man to love him. All guilt is he who eats with no partaker.[19]

The removal of hatred is a constant theme in the various appeals to the gods, not surprisingly, as hatred is divisive and thus works against harmony, *ṛta*. Even hate of one's foeman is considered wrong.[20] Deceit, lying stand against the very basis of the universe which is said to be 'truth' both in the Ṛgveda and the Atharvaveda.[21] In the latter is found an entreaty to be released from these:

> Man speaks much untruth, O King Varuṇa; free us from that distress.[22]

Deceit brings in its train all sorts of other evils that psychologically constrict the human psyche. Truth at all levels, speech, thought, action – human and godly – being the fundamental value, that which lies at the foundation of all, is conceived as sustaining all and through it alone all is secure, hence *ṛta*, the divine order. Soma, the god of rapture, flows

> declaring the law, splendid by law,
> declaring truth, truthful in action (IX.113.4ab).

Truth (*satya*) and law (*ṛta*) are in this verse almost synonymous, each emphasising one facet of Reality as viewed by human perception. Man's stand on truth, his inborn sense of fairness, justice, his acting and living in accordance with his 'truth', is his human expression of the great Law, Cosmic Order. So 'speak the truth and act the truth' is the epitome of Ṛgvedic ethics, an injunction which did not vary since we find it expressed in the Upaniṣads:

> Speak the truth, act the truth, never swerve from the path of truth (Tait. Up. 1.11.1).

> By truth is the divine path laid out by which sages, having obtained what they desire, ascend the supreme abode of truth (Muṇḍ. Up. 3.1.6).

> Truth alone triumphs and not untruth (ibid.).

This takes us back to the original Ṛgvedic conception, 'By truth is the world upheld'. This certainly gives the world the surest moral foundation. Man is a moral being, capable of choosing between right and wrong, and his choice of right will eventually make him like the gods who have aligned themselves on the side of right, thereby becoming fully *ṛta-visioned*.

Ṛta as truth

We have considered *ṛta* as Cosmic Order of which 'Truth' is the foundation. Some writers completely identify *ṛta* with truth at the expense of *satya*.

The modern analytical mind makes a difference between *law* and *truth*. It seems inconceivable that the two could be equated. Yet this aspect of the question of *ṛta* demands investigation. There are many kinds of law among humans, but ultimately there is only one Supreme Law which is Harmony, a constant, inner adjustment of all parts to achieve a balance, the mirror of Cosmic Harmony. Anything that contravenes that law is separative, conflicting, and breeds discord, suffering, evil. Harmony is right and therefore true; it is the divine truth; disharmony is wrong, and therefore false; it is a distortion of the divine Order. Law and truth find their meeting point in their ultimate basis, harmony. Truth is the basis of ideal Law, the celestial *ṛta*, and harmony is the basis of law and truth when viewed separately.

Lüders in his posthumously published work, *Varuṇa*, tried to prove that *ṛta* primarily meant 'truth' only,[23] and then secondarily 'cosmic power'; and that this secondary meaning developed as a result of the belief in the magical power of spoken truth, but as J. Gonda demonstrated in his review of Lüders' book[24] there is no such distinction of one meaning as against the other in the earliest strata of the Ṛgveda.

The 'act of truth', as appears already in the Ṛgveda, was indeed considered to be all powerful and capable of altering the apparent course of events, for it placed its enunciator at the very 'heart' of things. To the outsider it would look as if even the gods did bend to the will of the 'truth' speaker, whereas 'in truth' none acts, not even the gods, save in complete accordance with his innate tendencies, and destiny is already written within ourselves, however much outwardly we may seem to stray temporarily away from the intended goal; this because of those innate tendencies that only await favourable circumstances to be exteriorised and will inevitably bring us back to express what we inherently are. Because the 'truth' expressed by the speaker touches the very core of life, whatever be its outer aspect, whether he refers to the cosmos as in Ṛgv. X.37.2 or to himself, as e.g. the gambler's devastating compulsion which a complete facing and desperate acceptance is capable of cancelling out, he lays himself open to the working of the 'law' in a more dynamic way than ordinarily, aligns himself with it and thus becomes a co-worker with the law. Order is the truth of the universe and law its expression. *Ṛta* is not only the truth enunciated or 'breathed out' but the inner adjusting of all parts, the right but unseen layout which is a reflection, however limited, of the ideal cosmic order.

Hence the abode of *ṛta* is the centre point to which converge all forces, the point of perfect adjustment, harmony, the supreme heaven, whence 'they unharness the horses of Sūrya', that is, where all dissolves into perfect quiescence. It is this state that inspires the poet who, having tasted of it, or having permeated himself with it, emerges therefrom inspired and 'breathes out' the 'truth'. Hence *to vision the ṛta is to speak the truth*; truth, harmony, law are fundamentally one.

Sin as defilement

Vedic man did not escape from bodily ailments. Many hymns of the Atharvaveda are prayers for healing, exorcisms and spells. There we find personifications of what to the modern mind are abstract concepts such as fever, hostility or hatred, or appeals to Herbs, to Waters to cleanse and heal.[25] Ailments are usually considered the effects of sin so that sin is at times viewed as a pollution from which one asks to be cleansed. As explained by E. W. Hopkins, 'the sin causing divine anger has its outward manifestation in sickness or some other "lack of freedom" called a bond. Sickness is punishment for sin and is even the objectified form of sin.'[26]

The conception of sin as a defilement was to be developed in post Vedic ages, but from certain prayers it is obvious that sin was already occasionally so appraised. The Waters are the great purifiers. A remnant of this idea is still found at the entrance of Catholic churches in the holy water with which the visitor signs himself, thereby symbolically removing the 'pollution' or purifying himself from his sins.

> What impurity, what pollution, and what wrong we have committed, let the waters purify me from that.[27]

Soma and Rudra are asked to

> provide for our bodies all needful medicines to heal and cure us. Set free and draw away the sin committed which we have still inherent in our persons.[28]

The poet in the famous hymn to the herbs (Ṛgv. X.97) appeals to the plants to exercise their healing virtues and make a man whole, win back his very self. Disease of the body is an effect of psychological illness. The non-recognition of this fact led nineteenth century scholars to assess the Vedic ethics as purely materialistic or to see, as the great French scholar A. Bergaigne did, 'a confusion between the remedies of the body and those of the soul',[29] as though the two were radically separate, thus failing to grasp that the interaction between mind and body is so close that one – physical ailment – may very well be the effect of the other.

The ṛṣis' all unifying conception of Law and Order pervading all realms, cosmic, natural and human, is of the highest significance in the Vedic vision, for not only does it show a capacity to envisage an integrated cosmos but it also sets man, the disruptive element, right in the middle of the Order which he can only disturb at his own peril, and that, also, to a limited extent only.

EIGHT

Karma inherent in the conception of *ṛta* in the moral sphere

The doctrine of karma has unfortunately been all too often completely misunderstood or distorted beyond recognition, and blind prejudice even to this day has militated against its objective study and prevented scholars from observing that *ṛta*, the Cosmic Order, implies constant restoration of equilibrium at this phenomenal level of manifestation where the tension of the opposites is at its maximum – hence fundamentally the law of action and reaction, or karma.

For karma is fundamentally action and reaction, cause and effect, and therefore is inherent in the conception of *ṛta* the cosmic harmony. The root idea of karma which implies the setting right of any wrong action, the bringing back into harmony of what has become disharmonised, discordant, out of tune or out of order, is contained in the root idea of *ṛta*, in as much as *ṛta* stands for harmony and orderly process, that right working of all things, the inherent law of the universe which is the basis of manifestation; and in as much as *anṛta*, its opposite, is said to be punished by the gods who have taken their stand on the side of the cosmic order. The very mention that sin is punishable involves the notion of ill effects following wrong doing, of adverse reaction following wrong action. This is part of the idea of karma. It is never explained or hinted at in clear language in the Vedas, for, as we learn from the Bṛhadāraṇyaka Upaniṣad (3.2.13),[1] it was a secret doctrine. But as E. Hopkins rightly says

> though the Karma doctrine is not yet formulated, its ethical principles are already in evidence. Thus suffering is recognised as the fruit of previous sin and when a good man dies he goes to the next world carrying his merit with him.[2]

We should however try to eradicate our usual identification of karma with 'sin' only, sin conceived in a Christian sense. A wrong action is not necessarily sinful; to eat the wrong food which will impair the digestive system will necessarily have an adverse effect on the quality of the physical body, whether in one's self or in one's children (in the long run), and indeed on the life we lead, and thus a 'reaction', or 'karma'; and we may suffer for it, but this is not an 'ethical wrong' with which idea karma is usually and solely identified. On the other hand the very idea of suffering as a result of a sin committed implies retribution, an idea which is evidenced in the Vedas.

The significance of sin and its consequence, distress in some form or other, of reward of good action and punishment of evil doing, must have led inevitably to the philosophy of karma, or action and reaction. True, as already noted, we do not find any straight out injunction such as 'as ye sow so shall ye reap, as ye have sown so are ye reaping'; or as the clearly defined statement in e.g. the Bṛhadāraṇyaka Upaniṣad:

> As is his desire, such is his purpose; as is his purpose, such is the action he performs; what action (karma) he performs, that he procures for himself.[3]

He procures for himself the effect of his action. We are directly responsible to ourselves and our surroundings – which include the gods – for our every deed; this is expressed in the Vedas in various ways.

That the Adityas, undeceivable as they are, are credited with the ability to read the human heart, to look within, to perceive even when their eyelids are closed, beholding the crooked and the straight, observing the wrong doings of human beings; that Varuṇa has many spies that move among men and a thousand eyes, that numbered of him are the winkings of people; that the sun beholds the straight and crooked deeds of men;[4] that Soma

> parts like a knotted tangle right and wrong conduct[5]

– all these examples are the highly coloured Vedic version of the law of karma, of what H. Lefever refers to as a 'fact of experience, that every action of man, conscious or unconscious, produces its inevitable effect within the world order, ṛta.'[6] Such action is known to the gods and sweeps into its train a particular effect, be this termed the 'noose' of Varuṇa, should the cause be transgression, or sweetness, should the cause be right living. The verse

> Sinners are false and untruthful; they dig their own deep ditch[7]

is an epitome of one aspect of the law of action and reaction and a proof that the doctrine of karma was known to the ṛṣis even in its later emphasis on wrong doing followed by punishment.

Furthermore, as H. Lefever, in his study of Vedic sin, observes: 'when ṛta is disturbed, the results affect not only the sinner himself, but all men, for the disturbance is a cosmic one. Hence the frequent prayers that the Gods may not punish the sinner for another man's trespass.'[8] Pleas not to be punished for others are found:

> Let me not suffer, King, for what others have done.[9]

> Let us not suffer for another's sin, nor commit those deeds that ye, O Vasus, punish.[10]

If disturbances at the natural level affect all parties, so also at the human, ethical level. Hence the idea that the innocent suffer for the guilty. Not only is there interlinkedness among the gods but also among men, however men may ignore or deny the fact.

R. Shamasastry finds this suffering for others a proof against the prevalence of a belief in the doctrine of karma in

Ṛgvedic times. He takes a particular example: 'instead of ascribing the untimely death of a man to the sin he committed in his previous birth, a Vedic poet is seen to attribute it to the sin committed by his father.'[11] This is a very crude assessment, and indeed a distortion of the whole issue of the doctrine of karma and a baseless refutation of the idea that the doctrine may have been known in Vedic times. First, in the passage referred to, which is Ṛgv. VII.86.4 and 5, there is no 'untimely death' at all, but simply a question as to what transgression the poet could have committed that Varuṇa would slay him, his friend (if indeed the ascribing of such a wish to Varuṇa be not an exaggeration on the poet's part due to his repentant feeling or his distressed condition). The poet then implores:

> Release us from our fathers' offences, from those that we ourselves have committed.[12]

R. Shamasastry goes on: 'In the view of the Buddhists, the Jainas, and all the schools of Brahman philosophers each man is held responsible for his own Karma, and not for the Karma of his parents.'[13] That may be, but the author forgets that apart from individual karma the law includes group, national and racial karma, that we are all interdependent and jointly responsible to a degree for all that happens, an idea of which the ṛṣis seem to have been far more conscious than their descendants. The realisation of the interdependence of all, hence that the sin of one is also the sin of all, whether ancestors or contemporaries, seems to have been part of the Ṛgvedic doctrine.[14] To inherit the diseases of one's parents is itself an aspect of group karma completely missed by the learned author. The seed of diseases may have been sown by ancestors long ago through various abuses such as wrong eating or drinking or misuse of the physical body, and this is perpetuated in the diseases or handicaps of the descendants – the effects of causes sown in the past which cannot be easily overcome. (Witness children dying of delirium tremens at the age of twelve in France, for example, or children inheriting their parents' highly emotional or bad temper.)

Group karma, whether physical, or moral, or mental, is inherent to the human race.[15]

H. Lefever finds:

> Here the greatness and the weakness of the Ṛgvedic idea of sin become apparent. The conception of the universal world order checked an excessive individualism . . . it emphasised the responsibility of the whole human race for sin, and, furthermore, the part played by the 'unconscious'. These two important factors may tend to be overlooked in a religion based on the individual's relation to a personal God, who is Himself the ultimate source of the moral-religious Imperative.

This individualism has been pushed to such extremes in the West that hardly any Western scholar was able to appreciate the Vedic conception of interdependence, except H. Lefever. He, however, sees another side to it as well:

> On the other hand, just because of the lack of this ultimate individual relation to the ground of the Imperative, there is the danger in the religion of the Ṛgveda that sin may be regarded in too external a fashion. Though not the normal view in the Ṛgveda, there is a tendency to regard sin as a substance, endowed with a kind of *vertu propre* . . . which attacks and clings to a man like a disease and which can be removed by much the same method as in the case of a disease.[16]

This H. Lefever considers a 'material conception', whereas it could be merely a figure of speech. If sins were considered as the diseases of the mind or soul, the poets would easily choose their imagery from the material field, such as the waters that cleanse and thereby wash out the sins, or the medicinal herbs that purify. The conception of the remission of sin certainly degenerated into an external show of regular ablutions of the body taking precedence over the real inner cleansing of the heart. But such does not seem to have been prevalent in Ṛgvedic times.

Certain hints as to a current understanding of such a law, in the moral sphere, may be gathered from the word *ṛṇa*, 'debt', also used in the sense of *sin* in contexts which leave no doubt as to the meaning of moral debt incurred as an infringement against the divine law for which man has to pay the penalty. Similar hints may also be seen in a few enigmatic verses. Thus in Ṛgv. X.14.8 the departed is enjoined to

> unite with the reward of sacrifices-and-good-works

leaving his transgressions on the threshold of paradise (i.e. behind). The word translated as 'sacrifices-and-good-works'[17] is usually understood as the merits of sacred rites stored up in heaven, the rewards of the due performance of those rituals that the patriarchs instituted. It connotes the idea of duty carried out in the right, traditional manner. The effect of such duty will be rewarded in the next life. So the good that is done, or the 'well-made', usually refers to the right performance of the rite, hence the merit won through the sacrifice, but also to the acquitting oneself of all obligations, the fulfilling of one's duty. But in each case the cause begets an effect which is its exact counterpart.

A similar conception may be gathered from the Atharvaveda:

> Unite thyself with the fathers, with Yama, with thy sacred works in the highest firmament; abandoning what is reproachful, come again home; let him unite with a body, very splendid.[18]

What has been sown by way of 'good' is reaped as a reward in heaven to which the departed is said to come 'again'. What is reproachful is left as it were on the threshold; that, seemingly, will have its effect somewhere else, some other time. A curious question is asked in the Ṛgveda about a newly dead boy:

> How was his equipment

while the next stanza develops the thought further:

> According as the foundation was laid at the beginning of life so will the way out be contrived.[19]

This loose rendering of a rather enigmatic verse seems to mean that as it was at the beginning of life, so it is at the end; in other words, as we sow so we reap. The end of life and death will be the exact counterpart of how the life has been lived, a noble death is the expression of a noble life. This is one aspect of karma which tends to be overlooked but seems to be present in the Ṛgv. and the Atharvaveda.

On the other hand, those who do not live in accordance with the norms of the great Law will have to suffer the consequences of their having set up causes of disharmony, whether these effects be diseases which are the physical equivalent of moral wrong, or adverse circumstances; in breaching the divine order and its human reflection the social order, they bring about their own downfall:

> Retribution dogs the wrongdoings of men[20]

> Evil-doers do not cross the pathway of the Law.[21]

Whether the ill effects of wrong doing all manifest in this life, or in an intermediate state between heaven and hell, or in a future life, is not made clear. As already stated, the whole doctrine was secret. One may only surmise that what is sown at the subjective level, such as the emotional, mental or spiritual level, would be reaped at the subjective level, that is, in those states of consciousness that the ṛṣis hint at as prevailing after death; what is sown at the objective level, or physical life, would be reaped thereon either in this life or another.

The word *karma* which originally was *karman* has a long history with various connotations and emphasis, some of which may seem quite remote from what is now understood by it.

Dr R. Panikkar differentiates three trends of traditional meaning:

> *karman* as the saving sacrificial action, mainly stressed in the Saṃhitas; *karman* as the subtle structure of temporal reality, as that which all existing things have in common and in which they share, disclosed mainly in the

> Upaniṣads and developed in later times; *karman* as the path of action, of good works, and thus also as a way to salvation, emphasised in the Bhagavad Gita.[22]

Karman was originally the 'deed' *par excellence*, represented by the sacrificial act. The reciprocal gift from man to god and from god to man through the ritual is bound up with causes and effect, or the law of action and reaction, hence karma considered as ritual sacrifice. That which is offered up is bound to have a reaction on the offerer. The appeal to the god was considered to evoke a response from the latter. So Paul Yevtic opines that the later

> Karma concept as a law of retribution was only a modification of the earliest thinking – a gradual transformation of the idea which existed since earliest times. . . . There were many concepts underlying the conception of karma such as *yajña* (sacrifice), *dharma*, and especially *ṛta* – a universal world order in nature and ethical standards . . .[23]

Dr R. Panikkar sums up the meaning of karman in a way which bridges what may appear as a gap between this specifically early meaning of karman and the post-Vedic conception:

> We must stress . . . the sacrificial element implied within the concept of *karman* and not load it with ideas of morality which it does not primarily contain. The tendency to equate religion and morality, to see in *karman* a simple chalking-up of merits and demerits according to good or evil conduct, is secondary to authentic religiousness. For the Vedic Experience, religion is relationship of Man and the divine. *Karman* implies action . . . it implies an act of worship that is identified with sacrifice and worship.[24]

The stress is on karman as action, and that throughout the history of the word. The moral connotation is secondary to the fundamental significance of the word meaning 'action'

Whether life be interpreted as a sacrifice or not, its structure is based upon action and reaction at all levels, physical, emotional, ethical, intellectual. It is a constant take and give, an exchange, whether in growth, education, profession or relationship. He who cannot adjust in human relationship has failed to find harmony in himself and thus in the greater cosmic order. But to describe this 'exchange' and indeed the whole of the sacrificial ritual as a business deal between men and gods is but a reflection of our business mind and its incapacity to soar beyond the level of '*do ut des*'. The poetic mind sees here the eternal sacrificial life of Deity exemplified in the 'self-gift' of Deity's perfected creatures.

The Atharvaveda qualifies this law of cosmic order as 'mighty', or 'terrible', and brings together a number of epithets that reveal what lies at the basis of the universal order:

> Vast truth, mighty order, consecration, contemplative exertion (*tapas*), prayer (*brahma*), offering, uphold the earth.[25]

This implies that any trespassing of the Law, or right activity in its many ramifications, means discord and ultimately pain and evil, since the tendency is constantly to bring back the disorder into line with the order, and the friction thus created causes pain. Reaction restores the balance, hence *karma*. Therefore the closer human beings approach this fundamental equilibrium, the more does mankind become a dynamic expression of that truth which is the origin of all, the very pulse of being:

> Sweet blows the breeze for the righteous one, sweet flow the ocean-deeps.[26]

The Yajurveda adds:

> Where spiritual and ruling powers move together in unity, that world will I know as holy, where dwell the shining ones together with the Flame.[27]

The vision of holiness is here seen as the working together in harmony of all the forces of the universe.

All these examples show the conception of the law of action

and reaction or karma as already well developed. Just because a doctrine is called by a different name, and different aspects are emphasised in a different way, it does not mean that it was unknown before it assumed its later name. Nor was the later belief in karma such 'a moral advance on earlier ideas', as remarked by J. M. Farquhar[28] who thereby evidences his lack of insight into the Great Law, *ṛta*: the moral meaning, to use his own terms, which belief in karma gives to 'all conduct' was already present and developed in *ṛta*. On the other hand, to state as A. B. Keith does, that

> the idea of Ṛta is one which, like the moral elevation of Varuṇa has no future history in India, pointing irresistibly to the view that it was not an Indian creation, but an inheritance which did not long survive its new milieu.[29]

is to fail completely to assess the law of Cosmic Harmony, or of *karma* and of *dharma* in their essential significance. There is no gulf stretching between the Vedic *ṛta* and the Hindu *karma* and *dharma*; both the latter are the Hindu equivalent of the Ṛgvedic *ṛta*. *Ṛta* is more universal in its application and contains all norms, a cosmic order perhaps too vast for our puny human understanding. *Dharma* and *karma* in their later significance are more specialised and apply to humanity rather than to the cosmos. The vision of cosmic integrity, of which human integrity is the mirror and but an aspect, is peculiar to the Ṛgveda and is more and more adumbrated in post-Vedic literature.

NINE

Varuṇa and the fetters of sin

Of all the Vedic gods the great figure of Varuṇa stands nearest to the Christian conception of Deity and therefore is the one who attracted the most sympathetic attention among Christian scholars. The aweful majesty of Varuṇa is brought home to us again and again in some of the grandest verses of the Vedas:

> The wind thy breath hath sounded through the region
> like a wild beast that seeks his food in pastures.
> Within these two, exalted Earth and Heaven, O Varuṇa,
> are all the forms thou lovest (VII.87.2 Griffith's trans.).

Varuṇa's role as that of the controller and the binder is of paramount importance in the examination of the Vedic doctrine of sin. Being the ruler of the cosmic waters, that primordial substance out of which the universe is fashioned, he is described as the 'mysterious ocean', the 'support of the worlds' 'in whom is all wisdom'.[1] As he encompasses all creation he has control over all, he discerns truth and falsehood prevalent among men,[2] he 'binds' all and thereby brings back the sinner unto the right path, or restrains those who would stray away.[3]

Both he and Mitra, his constant companion, are described as peerless controllers, as 'equipped with reins', with their own arms checking the people or pulling them in, as it were with reins.[4] But they bind 'with bonds not made of cords'. Even all the gods follow the precepts of Varuṇa:

> This great god rules over all gods, for the precepts of Varuṇa the King are true (Athv. I.10.1).

> Before the home of Varuṇa all the gods follow his decree (Ṛgv. VIII.41.7).

The word Varuṇa is taken by most scholars to be derived from the root √vṛ, to cover, encompass.[5] This root also yields Vṛtra, the obstructor who also encompasses. The function of both these 'encompassers', Varuṇa and Vṛtra, seems at first glance similar; or, one may think that there is an overlapping of functions and characteristics, and one that somewhat tends to blur the outlines of each individuality; a reminder that the Vedic gods (and demons) are neither wholly individual nor wholly personal nor wholly impersonal.

But the profound difference between the two emerges on deeper scrutiny. Both Varuṇa and Vṛtra encompass, restrain, check, though each in a different way, the one putting a check upon transgressions, thereby in due course restoring harmony or order through his very checking and encompassing, the other obstructing the free flow of the divine life or waters of existence thereby causing friction and conflict. We are faced here with personifications of two relative forces of the universe that, in human eyes, work for the greater or lesser good. Yet it should be observed that restraint and expansion, friction and its surmounting, are needed in the universe and work in accordance with the law, so that the values of good and bad attributed to these forces are, in the last analysis, purely human and subjective.

Varuṇa makes the claim that he rules 'over man's highest vesture'.[6] The keyword translated as 'vesture' refers to a highest place of concealment or covering of man over which Varuṇa is king, which may be not only 'heaven' but what was later to be explained as 'sheath', 'covering' (*kośa*) in the fourfold division of the human constitution of the Upaniṣads. The very word 'highest' in the Ṛgvedic verse hints that there must be several such 'coverings'.[7] The fact that Varuṇa rules over this highest vesture might mean that his is the last veil that conceals the innermost core of the human being. Varuṇa not only encompasses the universe but also man whom he covers with his supreme veil, the purpose being to shelter man from the too great effulgence that lies in his innermost core and to which he refers as the 'sun'.[8]

On the other hand, Vṛtra holds back the waters of life or hides the sun, and Vala holds hidden the cows or rays of light which the warrior god Indra releases after much effort. In the one case we glimpse the protective end of covering, in the other the inertia (*tamas*) of nature – obstruction, friction, darkness – which, translated in mythological terms, becomes the selfishness of demons.

Varuṇa not only covers but he also fetters. His fetter or noose (*pāśa*) is dreaded by the sinner who implores to be released from it.[9] The latter is an image expressive of the effect of the disturbance of *ṛta*.[10] The sinner is enchained by his own sin, he has transgressed against the laws of the sons of freedom. In human eyes, the great Aditya, Varuṇa, son of freedom (*aditi*) punishes by fettering, i.e. the effect of going against the law which he embodies results in man's having to suffer the opposite of the god's essence which is freedom. In taking up his stand against established law, the cosmic order, man binds himself to disorder, chooses bondage as against freedom; to bind oneself to freedom means to free oneself. In Varuṇa we see the expression of two possible ways which describe man's choice: freeing and fettering – hence his apparent 'dark' as well as 'light' side – only in regard to man's evaluation of light and dark, good and evil, for human reaction is always to blame the other party; so the Adityas are said to deceive the wicked and the unwary who are blind to their own misdeeds and weaknesses and, when punished and brought to their senses, accuse the gods of having deceived them! The bonds of *anṛta*, it is said, are 'equipped with many nooses' 'difficult to overcome' for 'the evil mortal' (VII.65.3ab).

Retribution comes in the shape of 'nooses' or 'avengers':

> At every step are bonds equipped with nooses (IV.73.4d)
>
> Avengers follow the falsehoods of men (VII.61.5c).

In two passages the noose itself is described as triple, though the meaning of this threefold bond is not explained:

> Untie the uppermost fetter from us O Varuṇa, the lowest, the midmost; may we then O Aditya, in accordance with thine ordinance belong, in all purity, to Aditi (I.24.15). Release us from the upper noose, loosen us from the midmost and the lowest, that we may live (I.25.21).

Just as the cosmos, in the Vedic conception, is threefold – the earth, the intermediate region and the heaven – so is the microcosm, man – physical, psycho-mental and spiritual; so the threefold bond might apply to an analogous division in each of which man may fetter himself through his own action, the fettering at one level reacting upon the other levels. But the fetters may also be an image of the 'covering' which hides the next level of reality from human view. Only when these three coverings are lifted through righteous living, when these three fetters are untied, can human vision be pure, single, whole, can 'we, in all purity belong to Aditi', i.e. be free, claim our heritage, freedom.

The fetters of Varuṇa, described as such because typical of the effect of transgression against Varuṇa's statutes, are both imposed upon humans as well as unclasped not only by Mitra and Varuṇa but also by other deities.[11] Agni is asked to

> loose from us the bonds that bind us (V.2.7d).

Rudra and Soma, the former noted for his deadly missiles, are begged to

> be merciful to us, release us from the noose of Varuṇa, in your graciousness protect us (VI.74.4bcd).

But in reality none can escape from Varuṇa's eye:

> There is no escape from King Varuṇa (even) for whosoever would creep further away from the sky (Athv. IV.16.4).

He stands undeceivable and is implored not to strike mortals with his dread weapons that wound the sinner:

> Strike us not, Varuṇa, with those dread weapons which, Asura, at thy bidding wound the sinner.[12]

Most hymns addressed to Varuṇa are prayers showing a recognition of sin, a dread of Varuṇa's punishment, of his 'wrath' which he seems to manifest often, and repentance and aspiration to be made whole:

> If in our wantonness we have violated whatever-you-have-laid-down-as-law (*dharma*), for that offence, O God, do not harm us (VII.89.5cd).

The harm that the sinner feels is inflicted upon him by the god is but the effect of his sin. But what of the wrath of Varuṇa? What is the meaning of anger in a deity?

The wrath of God is a notable feature of Deity in all the scriptures. In the Vedas Varuṇa and Rudra have a good dose of this vehemence. To the rational mind the ambivalence of Varuṇa is difficult to understand. That a deity as great as Varuṇa, the Lord of the Cosmic Order, seems to have been in the Vedic pantheon should contain the light and the dark side of the universe so prominently and act in strange and at times devious ways, does not appeal to our sense of righteousness. Yet, that sense cannot answer the question 'whence the darkness and why and what is its purpose?' Varuṇa is as full an expression as the human mind can create in terms of a more or less anthropomorphic Deity, of that experience of the numinous which R. Otto tried to analyse and explain in his deep study of *The idea of the holy*.

The question that is posed when we consider the darker side of deity may be summed up thus: how can God be offended, wrathful, jealous, indignant, vehement, vengeful? These seem human judgments passed on the deity, but ones which, notwithstanding the rational mind's inability to accept these, hide profound truths of the human psyche and its reaction to its deepest, darksome side.

That the wrong-doer is pursued by the voice of conscience or guilt feelings which, under certain conditions, may culminate in a complete breakdown or warping of the personality; that the man who turns his back on his inner self, scorns, ignores, represses the pull of that half of himself that demands recognition and is the seat of so much inspiration and spiritual insight as well as the storehouse of man's past both good and evil, both light and dark, risks his eventual undoing, for it will relentlessly pursue him like Artemis, the divine huntress with her deadly hounds, until he too falls in the maelstrom of madness – these well known examples of the human plight and the human struggle are summed up in these ancient expressions found in all the scriptures of the world, the 'jealousy', the 'vengeance', the 'wrath' of God:

> For I the Lord thy God am a jealous God (Exodus 20:5).

It is the struggle between the two halves of human nature, the unconscious and the conscious, the spiritual and the material, the male and female, that is depicted in the imagery of wrath, jealousy, vengeance, and it is also the capacity to submit, to accept, to integrate that is shown as the mercy, compassion, forgiveness of God. Both these aspects are present in the following address to Rudra:

> May he the compassionate, the easy of invocation, spare us his jealousy.
>
> May the dart of Rudra pass us by, may the great ill-will of the terrible one go by us (II.33.5d, 14ab)

for Rudra's shafts that slay like a dread beast are terrible, but his hands dispense, as well, wholesome, beneficent remedies. Equally is Varuṇa the god of wrath and graciousness.

The strange, non-rational aspect of divine visitations – what the Greeks called theophany – as experienced in certain encounters with the numinous, the filling of the devotee with sudden awe, dread, terror, at contact with the uncanny side of the divine influx, his feeling utterly 'unworthy' as a creature and even 'guilty', account for that most extraordinary infliction of irrational traits upon the Deity such as the 'wrath of God', his jealousy, fury, fierceness, vengeance. Such anthropomorphic projections of the irrational side of human nature abound in the Ṛgveda, the Purāṇas, the Old Testament, the New Testament. No wonder that at contact with that all-compelling energy akin to all burning love, that flame-divine that sweeps all forward, that is 'dread-full', devastating and relentless, the worshipper ascribes to his god wrath, fury and so on. The Israelites constantly refer to the 'wrath of Yahweh' and St Paul exclaims

> It is a fearful thing to fall into the hands of the living God (Hebrews 10:31)

> For the wrath of God is revealed from heaven against all ungodliness and unrighteousness of men, who hold the truth in unrighteousness (Romans 1:18).[13]

Compare the Ṛgveda: 'avengers follow the falsehoods of men' (VII.61.5c). As R. Otto explains:

> 'Wrath' here is the 'ideogram' of a unique emotional moment in religious experience, a moment whose singularly daunting and awe-inspiring character must be gravely disturbing to those persons who will recognise nothing in the divine nature but goodness, gentleness, love, and a sort of confidential intimacy, in a word, only those aspects of God which turn towards the world of men.
>
> This *orge* (wrath) is thus quite wrongly spoken of as 'natural' wrath; rather it is an entirely non- or supernatural, i.e. numinous, quality. . . . Something supra-rational throbs and gleams, palpable and visible, in the 'wrath of God', prompting to a sense of 'terror' that no 'natural' anger can arouse.[14]

The fury of Varuṇa's hostile wrath that fatally slays the transgressor[15] is obviously one such experience of falling into the terror of 'the living God':

> Whatever ordinance of thine, do we, as men, daily transgress, O God Varuṇa, deliver us not to the fatal slayer, to the fury of thine hostile wrath.[16]

Varuṇa's fetter seems to entail death or disease. Thus the poet entreats:

> Be thou not displeased O Varuṇa; attend thou here, O thou praised-by-many; Do not steal away from us our span of life.[17]

'Wrath' in the Vedas is personified as *manyu* to whom two hymns are addressed.[18] Glowing like fire to which it is closely akin, compared to Indra's destructive thunderbolt of which it has the killing power, this wrath, Manyu, is the companion of the storm gods, the Maruts, but is also said to be Indra, Varuṇa, Agni, so evidently is an attribute of these.

Manyu is a potency, a raging fierceness that gives the warrior superhuman power, but a potency accordant with that ardour or flame power (*tapas*) by means of which the contemplation of the Supreme bore fruit and the universe was manifested.[19] We touch here certain numinous qualities of the fire in both its creative and destructive energetic aspects, twin powers of the one all blazing Energy which Vedic man distinguished as *tapas*, creative fervour, and *manyu*, destructive potency, yet creative when allied to the former, of which certain powerful radiances such as *tejas* and vital energy *ojas* are the offshoot.

The energy that is represented by *manyu* is said to slay enemies, Vṛtra, and so on, and is invoked for battle and protects the worshipper:

> Come hither Manyu mightier than the mighty; chase with thy fervour (*tapas*) for ally, our foeman.
>
> For thou art Manyu of surpassing vigour, fierce, queller of the foe and self-existent. Shared by all men, victorious, subduer; vouchsafe to us superior strength in battles.[20]

When it allies itself to mental power (*kratu*) or skilfulness (*dakṣa*) it becomes an irresistible creative agency which distinguishes the great gods such as Mitra-Varuṇa. By that fiery energy of wrath as it worked upon the skill of Mitra and Varuṇa, by that blazing one-pointedness, the latter are said to have sorted out the cosmos from the chaos, the order from the disorder, the falsehood from the true (I.139.2). But *manyu* as sheer anger, i.e. uncontrolled, can lead astray, as the poet's confession to Varuṇa admits:

> Not our own will betrayed us, but seduction,
> thoughtlessness O Varuṇa, wine, dice or anger!
> (VII.86.6).

Dice elsewhere is mentioned as the only vice that does not bow before the wrath of the gods!

The fire of wrath can be appeased by the fire. Since Agni is the 'knower of Varuṇa', it is trusted that he will appease the latter's anger and 'release from us all hatred'. The freeing from all types of hatred credits Agni with the power to release the human being from impurities which arouse Varuṇa's anger, but can be removed either through Agni's intercession (by way of the sacrificial ritual) or through Varuṇa's mercy, as the case might be. So the fire protects the devotee from the fire of Varuṇa's wrath, 'from the great god's chastisement', or harm (I.128.7d).

Varuṇa's anger may also be appeased through righteous conduct, through the chanting of his praises and through the ritual, through sinlessness, through being his 'slave'. Ever the refrain is 'Varuṇa . . . be not angry', 'we turn aside thine anger':

> Like a slave will I render service, sinless, to the bounteous, angry god. The noble god made the thoughtless think; he, wiser, speeds the experienced to riches (VII.86.7).

But Varuṇa is not merely the Lord who chastises, he is also the Lord who shows mercy to the repentant sinner, the one who delivers from the fetter of sin, the wise one who stands nearby in times of stress, the friend to whom one may talk – in spite of his majesty.[21] To Varuṇa pardon is asked for trespasses whether committed by oneself or by others, his favour is entreated to loosen the sinner from his sin as from a rope,[22] that men may succeed in finding the 'aperture of *ṛta*',[23] hence open their consciousness to the significance and implications of the divine order and thereby live accordingly. The wish is expressed often enough:

> O that we be found guiltless in the presence of Varuṇa who is merciful even to him who commits sin, as we accomplish Aditi's laws (VII.87.7abc).

Deliverance from wrong-doing is ascribed to Varuṇa as though to his compassion:

> Our wrong-doing which faultless Varuṇa perceived, the wise one has certainly delivered us from it (VII.28.4cd).

Varuṇa's healing remedies are countless; they are the gift of *life*. He is asked to 'prolong our lives for us'. There is always hope, through prayer and offering, that his wrath should subside, that his

> good disposition should spread far and penetrate deep.[24]

The words 'good disposition' (*sumati*) and 'favour' (*mṛlīka*) or mercy are distinctive epithets of Varuṇa. He is the great Lord before whom the penitent is powerless, yet keeps hoping. This is summed up in the following:

> O Varuṇa keep dread far away from me; O Sovereign, holy Lord, be favourable to me. Like cords from a calf, release this distress from me. Without thee I am not even lord of my eye's winking! (II.28.6).

We are entirely in the Lord's hand!

Forgiveness and retribution, opposites though they may appear, find their resolution in human change of heart.

TEN

Forgiveness of sin and the meaning of freedom

That human nature is liable to transgress, to err widely, that at times man does not even realise that he has committed wrong and therefore begs for the gods' compassion and understanding of human weakness, these ideas are expressed again and again:

> If we have erred intentionally, unintentionally, and by way of imprecation, waking or sleeping, may Agni remove far away from us all evil and disgusting actions.[1]

The belief that retribution follows sin, that wrong committed will have to be paid for in some unpleasant way or other, can be further inferred from the many entreaties for forgiveness addressed not only to Varuṇa but to other gods. Two trends of thought are clearly perceptible, that of retribution and that of forgiveness, for the conception of the remission of sin is also evident; the god, e.g. Varuṇa, may forgive out of his own graciousness, sin can be wiped out; or else, as said of Brahmaṇaspati, the god may remit human debts at his will. The power to bind and unbind sinners, that is, to punish or forgive, thus rests with the gods. They are, in human eyes, the agents of the law. The many pleas to be 'gracious' or to show mercy (as the verb *mṛl* indicates) imply that the repentant sinner could expect forgiveness on the part of his god even though he had infringed the law and even though the Adityas are its custodians. The verb 'to be gracious' (√*mṛl*) and the substantive 'mercy' (*mṛlika*) are very often used with reference to Rudra, Varuṇa, Soma, thus showing a belief in the possibility of sin's annulment or the wiping out of human 'debt' through the deity's compassion or will. As pointed out by Dr Panikkar, 'mercy and compassion have their place in the framework of the universe',[2] for the universe does not work on mechanical, automatic laws.

The key to the apparent break of divine statute, or of the law of cause and effect, must be found in the attitude of repentance, or rather conscious and total change of mind and heart on the part of the sinner which would transform his whole being and give him a new lease of life, hence in his eyes 'forgiveness' in the sight of the gods, since it is human to externalise or project objectively that which is basically subjective. The realisation in depth by a human being of what he has done changes his ways, and thereby effects a transformation in his life which he ascribes to the gods' forgiveness. The Ṛgveda abounds in such appeals for forgiveness.[3]

Dr R. Panikkar explains this idea of mercy

> as the fruit of a relationship, but the relationship is not exceptional nor does it militate against existing regulations; it is a part of the overall order of *ṛta*. Neither *ṛta* in the first period nor *karman* as it slowly begins to emerge in the pre-Upaniṣadic period is an automatic, merely mechanical force; both *ṛta* and *karman* are always functional and they function according to a set of relational factors, one of which is the human will along with its sentiments and feelings.[4]

Such an idea seems to be touched upon in both Ṛgv. VIII.67.17 and II.24.6 where we find the Sanskrit participle (*pratiyantam*[5] in the first example) bearing the meaning of 'turning back' upon oneself, thus transforming one's view, pointing to the Vedic seer's realisation of the necessity of a change of mind and heart in order to be 'forgiven':

> To everyone turning back even from sin, O ye wise gods, ye make to live (VIII.67.17)

or perhaps 'ye grant a further lease of life'. As a result of this re-direction of consciousness, the gods 'make to live' or renew life in a deeper sense than the merely physical renewal. This is the complete psychological turning round that makes for transformation from the roots. E. W. Hopkins, taking a different example (Ṛgv. X.71.6), explains this 'turning round' thus:

> one who betrays his friend loses his hold of the word 'because he has no understanding of the paths of well-doing'. This is the path to which one returns when one becomes 'sinless before Aditi' or in particular before Varuṇa, whose laws determine the straight course of the stars. . . . Only when one has 'returned' can one be forgiven: 'ye make to live, O ancient gods, every one who returns from his sin'. This return is implied in petitions where the suffering sinner begs for relief.[6]

Renewal from the depth through this complete turning is forgiveness.

A somewhat similar turning back which, in the particular case to be considered, is a return to the source, may be glimpsed in the more enigmatic verse of Ṛgv. II.24.6 where the words 'placed in the cave', i.e. hidden away (*guhā hitam*), and 'that' (*tat*) are of deep significance, the former always pointing to the seat of the heart where is hidden the greatest treasure[7] and the latter pointing to the essence of Deity with which the Upaniṣads were to equate the human spirit.

> They drew near; they reached out towards that supreme treasure of the Paṇis secretly hidden away: those knowing ones, having viewed the falsehoods, turned back again whence (they had come) seeking to penetrate *that* (II.24.6).

That which the sages wish to enter is the 'supreme domain', the state of immortality which is the highest treasure kept hidden away in darkness.

The same idea of observing the world of falsehood, seeing through it, rejecting it and turning back, is repeated in the following verse and the sages are then declared to ascend the high paths:

> The law-abiding sages, having observed the world of falsehood thence (turned back) again and stepped unto the high paths (II.24.7ab).

It seems that if there is no turning back, leading to a new start, but only arrogance, then the god does not forgive; e.g. Indra 'forgives not the arrogant man his arrogance' (II.12.10c), forgiveness being contingent upon a change of mind and heart.

The forgiveness of sin is conceived in terms of a removing of defilement, but especially as a 'release', or 'untying' from 'bond' and therefore a freeing.[8] Rudra is asked to

> drive away from us hatred, away distress, away diseases in all directions (II.33.2cd)

all of which – apart from hatred which can itself be a cause of sin – are the results of sin. The Waters are implored for their cleansing effect, theirs being a purifying function:

> Whatever sin is found in me, whatever evil I have wrought, if I have lied or falsely sworn, Waters, remove that from me (I.23.22).

To Agni the prayer goes:

> O youngest one, whatever sin we have committed, through our human condition, through want of sense, make us sinless in the sight of Aditi, pardon every offence whatsoever, O Agni.[9]

That even great sin could be forgiven is indicated in the following verse:

> Even before great sin, O Agni, (free us) from prison of the gods and mortals.[10]

The list of pleas for forgiveness is rather long, many of which are not addressed to Varuṇa only, but also to other gods, to Heaven and Earth, to Dawn, Rudra, Indra, Savitṛ. They certainly show the Vedic consciousness of the gravity of transgression against the law and the all too human fear of the consequences.[11]

> Not for one offence (*āgas*) O valiant one, not for two, nor for three, do thou strike us, nor yet for many (VIII.45.34).

The common plea is that whatever be the offence which we, human beings, in our wantonness (*acitti*) commit against the heavenly hosts, whether through lack of intelligence and understanding, or human weakness or sheer wickedness, 'do not hurt us'.

The fact that sin is unacceptable in itself and not simply because it involves retribution is also evident:

> If we have ever committed sin against a loving one, a friendly one, companion or brother, the neighbour always with us or a stranger, do thou O Varuṇa loosen that from us (V.85.7).

Many an entreaty to Aditi is for forgiveness of sin.[12] Similar entreaties are addressed to other gods that man be made 'sinless in the sight of Aditi!' or 'unto freedom', sinlessness being a condition of freedom and safety:

> Whatever sin we have committed, that forgive us, Aditi, Mitra, Varuṇa,
> May I obtain freedom and light free from fear, O Indra, may the spreading darkness not reach us (II.27.14).

A similar request is formulated thus:

> May I attain the fearless light (II.27.11d)

obviously that light in whose radiance man can have no fear. As in the first example the desire evidenced for freedom and light follows upon the plea for forgiveness, we surmise that this freedom is the state of purity, a cleansing from trespasses which otherwise lead to distress, guilt, moral bondage and physical disease: complete harmony with *ṛta* meaning freedom, hence salvation.

Aditi the luminous one, the upholder of all creatures, the holy, and her sons protect or rescue human beings from guilt, anguish (*aṃhas*), from every kind of evil or malignity, these in human terms of reference, being connected with disease, want, obstruction and thus destruction or disintegration (*nirṛti*) and death:

> May Aditi protect us, may Aditi grant us shelter (VIII.47.9ab).

The boons of the Adityas are the opposite of those evils that beset the unrighteous, they are beneficent (*avadha*), not hurting, two of their greatest gifts being 'everlasting light' (*jyotir-ajasram*) and 'wide space' (*varivas*), i.e. free scope or freedom from anguish and all kinds of distress; hence the oft-mentioned opposition between wideness, which implies freedom of range and expansion, and distress, anguish, which implies constriction, lack of scope. Expansion is inclusive, and therefore leads to wholeness; constriction is exclusive and therefore leads to separation, so that we have in these two ideas the fundamental dichotomies of love and hatred, harmony and conflict, light and darkness, right and wrong. So to the gods the poet confesses that away from constriction (*aṃhas*) there is spaciousness (*uru*), hence freedom; the whole world belongs to the pure:

> O gods, O sons of freedom, O ye that transcend all evil, from distress there is freedom, for the sinless a treasure.[13]

Brahmaṇaspati also grants his worshipper relief from distress, literally grants him broadness out of narrowness:

> He safeguards him from distress, defends him from injury, granting him ample and marvellous protection against distress (II.26.4cd).

Similarly

> Neither distress nor woe, neither evil nor dishonest men from any side whatsoever overcome him. Thou drivest away all noxious powers from that man whom O Brahmaṇaspati thou protectest as a good herdsman (II.23.5).

> Now Mitra, Varuṇa, Aryaman vouchsafe us freedom (*varivas*) for us and for our offspring (VII.63.6ab).

The 'ever-strengthening' (*sadavṛdha*) Aditi protects men from coming to grief, that is from *aṃhas*. To find refuge in the Adityas means release from bondage, what in later times was to be called liberation (*mokṣa*):

> O bounteous Adityas vouchsafe us that refuge that releases even the sinner from his sin (VIII.18.12).

So Aditi, Mitra, Varuṇa, Indra, are begged for

> forgiveness for whatever sin we have committed.
> That I may obtain the broad, the fearless light.
> O Indra let not the long-enduring darkness overwhelm us.[14]

Light cannot be qualified as 'fearless'. The poet is evidently transferring a certain psychological experience he may have had, an enlightenment marked by freedom, joy and fearlessness, to the manifestation of light or illumination itself. This broad fearless light which is to be obtained after due forgiveness of one's transgression is described as celestial (*svarvat*) and is the boon begged of Indra:

> Lead us to wide space, O thou who knowest, to celestial, fearless light, successfully (VI.47.8ab).

The fearless light could also refer to freedom from the shackles of mortality, that wide place, *urum lokam*, or state of being free from all bondage.

Forgiveness which implies a new state of mind and heart lifts the darkness from the human mind thus renewed. B. G. Tilak was inclined to take literally those references to 'long darkness' which we find scattered in the Vedic hymns. Around these he built his theory of the Arctic home of the Vedas, such references being reminiscent of the long winter darkness peculiar to the polar hemisphere where, according to him, the Aryans lived before they settled into India.[15]

In all the scriptures of the world 'darkness' plays a vital part, not only cosmically, as the matrix of the universe whereon breathes the spirit of God and wherefrom all emerges, not only as the dark womb of earth out of which grows the seed to maturity, but also as that psychological darkness that, on the one hand, prevents man from 'seeing' and 'understanding' as a result of which he errs, and, on the other, represents that unconscious that holds hidden in its secret recesses the pearl of great price that is man's task to find. A hint of this is given in the Old Testament:

> And I will give thee the treasures of darkness and the hidden riches of secret places (Isaiah 45:3).

The 'long darkness' that the Vedic bard fears (II.27.14) is, we surmise, more of a psychological darkness, a fear of not being in the light, enlightened, as is evidenced in another prayer 'let us not go as exiles from the light' (II.28.7c), light being the great boon yearned for by the bards, that 'heavenly light' (*svarvaj jyotiḥ*) which the sages long to obtain (X.36.3); that 'sun' that is discovered as 'the loftiest light beyond the darkness' (I.50.10) which 'mortals behold not' (I.105.16) is the *ātman*, i.e. the *self*, the innermost essence 'of what moves and what moves not' (I.115.1), hence not the sun in the sky, though as effulgent as that, but the light that radiates from all beings and all things, 'that eternal light within all beings without which nothing can be made' (Yjv. 34.3). It shines beyond the darkness of the senses and the mind and is the *puruṣa*, the Heavenly Man, on knowing whom one overcomes death:

> I have known this Heavenly Man, refulgent as the sun beyond the darkness;
> Only by knowing him does one overcome death; no other way is there to go (Yjv. 31.18).

The opposite of this light, the obscurity that hides it, causes fear in the poet's heart; this darkness depicts the state of sin and/or ignorance, of searching gropingly for truth, for understanding, hence the prayer to the Aśvins to 'vouchsafe such strength to the worshipper as may carry him through the darkness' (I.46.6) and those prayers for 'sight' which may not be merely physical, but also spiritual sight (X.158.3 and

4). For we, as humans, 'see through a glass darkly' as it were, and cannot tell when we shall see clearly. So the bard of the Atharvaveda invokes night as 'mother' asking her to commit us to the dawn:

> Each moving thing finds rest in her (night) whose yonder boundary is not seen nor that which keeps her separate. O spacious-darksome night, may we, uninjured, reach the end of thee, reach O thou blessed one, thine end! (Athv. XIX.47.2).[16]

This does not necessarily refer to the long Arctic night but to the night of the soul when it has lost its vision and knows not where to turn, though it still knows in an obscure way that within the darkness is hidden the eternal light that will give it freedom. In another verse, the bard refers to night as 'profound one', 'spreading towards me like a friend' (Athv. XIX.49.2).

Freedom is a recurring theme in Vedic prayers, freedom from all limitations, from all that which hampers the free flow of life, that freedom that removes all obstacles, physical, moral, mental, all transgressions against the law, all diseases, poverty or oppression, and darkness; that freedom whose basis is purity, harmony, well-being. The sacrificial offering is to be placed in a state of sinlessness, that is purity, wholeness, as well as freedom. Complete purity means wholeness. When all the body's functions work normally and therefore in harmony, the man enjoys good health; he is 'hale', healthy, whole, at the physical level. When his mental faculties also work well and in harmony, and his emotions are under full control, he is a harmonised man, he is whole. Only a fully harmonised man can be a mirror of *ṛta*, the divine harmony, and perform his proper function in accordance with the divine decrees. So we find the prayer:

> May we be free, (may we be) sons of freedom, (may we be) a stronghold midst gods and mortals, O ye Vasus (VII.52.1ab).

To Agni, the divine flame who is identified with 'freedom', *aditi*, the bard says

> thou grantest sinlessness, freedom, completely (I.94.15ab).

The word translated 'completely' (*sarvatātā*) recurs in its adjectival form in the refrain of a hymn addressed to all the gods (X.100) asking for the boon of freedom and completeness, which could be rendered as: 'We choose freedom and wholeness'. This points to the total integration of the human being – an aspiration, certainly, among the seers, and perhaps occasionally an achievement![17]

We may have in the contrast between 'anguish', constriction (*aṃhas*) and 'space', free scope (*varivas*) or freedom (*aditi*) a glimpse into the Vedic conception of sin. The physical condition of anguish is used as a vivid, metaphoric image to describe the psychological state of sin: sin is constricting, darkening, in some subtle sense it means imprisonment, hence the 'fettering' of the human being in so far as his choice of right action is concerned; by 'fettering' himself he has forfeited his right to choose, to know the truth, to be 'free'. Righteousness implies expansion to the light, hence space, broadness, freedom, hence the many aspirations to the light. It is an opening to a vaster dimension, a harmony of one's whole being with the cosmic *ṛta*, an integration in the universal harmony. It is *truth*. This alone can lead to wholeness. Truth, integrity, in the human being, are expressions of his wholeness. They are the moral equivalent of harmony. They are *ṛta* as it applies to human norms. Hence the prayers for freedom and light. Light means enlightenment. The latter grants supreme understanding. This conduces to right action, hence harmony, integration in the vast whole.

> May we attain that most spacious mansion, your own dominion (V.66.6)

seems an epitome of the bard's yearning for freedom.

Moreover, the gift of freedom, the great boon of Aditi, the 'incomparable share' which the poet claims from Aditi seems to be Aditi's light; on this 'beneficent light he would meditate'.[18]

This light cannot be obstructed, over it the evil-minded has no power:

> the enemy has no hold over those to whom Aditi's sons give their eternal light to live.[19]

To be made sinless in the sight of Aditi (IV.12.4c) may thus mean to be so purified as to be able to enter that light. It is evident that those who live in accordance with the light of the gods, who follow their ordinances in the harmony of the *ṛta*, can fear no adverse consequences:

> For the righteous one sweet blows the breeze,
> sweet flow the oceans (I.90.6ab).

As pointed out by H. Lefever, 'prosperity is the reward but not the goal of right-doing'.[20]

So the gods are implored to

> transport us to the farther-shore, away from distress to safety.[21]

Trust in the gods and a willingness to abide by the path of *ṛta* as outlined by them for men is the best safeguard. The prayers, of which a few samples are given below, show man's trusting relationship with his god:

> By means of your path of *ṛta*, Mitra, Varuṇa, let us cross over the (ocean of) evils as a ship over the waters.[22]

Agni, the 'knower of births' is entreated to

> lead us across the ocean of all difficulties, even as a ship across (all) perils.[23]

A similar prayer goes to the Vasus:

> As a chariot (is brought) across a difficult passage, so do ye O bounteous Vasus, deliver us from all distress.[24]

For 'the paths of *ṛta* are fair to tread'[25] for those whom the gods protect from all kinds of evil, those who remain true to the gods.

> Whoever is in thy grace is not slain, is not conquered,

> him distress does not reach, either from far or from near.[26]

This verse bears witness to the ṛṣi's faith and trust in his god.

ELEVEN

Ṛta as truth: the Vedic vision of holiness, truth, and the social order

The great unitary vision of Vedic man made him capable of conceiving a oneness to which all aspects converged: all things working towards a greater harmony of the whole so that whatever brought disharmony was 'wrong' and in man's case 'sin'. Dr R. Panikkar expresses this thus:

> The dichotomy between an ethical and a cosmic order is foreign to Vedic thinking, not because the ethical order is ignored but because the really existential order is anthropocosmic and thus includes both the ethical and the cosmic in one.[1]

Ultimate differentiations of the extreme analytical kind are not typical of the mystical viewpoint that embraces all aspects as a whole; the ethical order pertains to humanity and humanity is part of the cosmic order, hence the use of the adjective 'anthropocosmic'. At our human level the one law of harmony finds expression in human righteousness and social order. Hence the laying down of social and ethical norms for man to follow so that as an individual and as a group, i.e. in society, man can endeavour to work towards the end of establishing on earth the ideal cosmic order. Set in the middle of this order, man is shown as capable of becoming an active co-worker with the agents of the cosmos, the gods, to bring about in his own society the harmony that prevails in the 'heavens'. In the task of actualising this at the phenomenal level, the gods and man are shown as closely linked and able to work together, both through their right attitude and right deeds and through the ritual which acts as a bridging between both parties. The solidarity which we have seen prevalent among the gods is an ideal set before humanity to work upon.

We can hardly agree with A. C. Clayton's claim that 'the supreme vision of holiness was simply not attained'.[2] This is a Christian prejudice capable of seeing 'holiness' only in terms of an anthropomorphic, ethical Deity whose only Son is sent to redeem humanity through his sacrificial death, an offering of the first and best fruit. In thus ascribing to the word 'holy' what R. Otto describes as 'an entirely derivative sense',[3] namely the consummation of moral goodness, quite different from the original meaning which, although it included the moral quality, went far beyond it in its essential significance, the 'overplus' of meaning, is forgotten; but it is still present in the cognate old English 'hal', giving 'hale'. Hale and whole, without flaw, hence entirely good, hence inviolate, point not only to 'wholeness' as being part of the essential meaning of holiness but also to a surplus or substratum of something 'other' than the well 'known'; for we know 'in part', 'darkly', which is different from knowing as a whole; this *otherness* is implied in the word numinous coined by R. Otto. The experience of the numinous is the mystical experience of the divine, which is quite apart from any category of 'good'.

In so far as the Vedic vision of the cosmos is concerned, the sages show a remarkable capacity for an all-comprehensive view of the universe and life; this is envisaged in terms of those elements of majesty, overpoweringness and awe-inspiring sublimity that mark their vision out as supremely 'holy'.

The vision of the Cosmic Order born of the all-blazing contemplation of the Divine Overseer, and of the infinite variety of Deity's manifestations – both in terms of divine agents such as the 'gods' working together[4] towards the greater end of establishing on earth the harmony that is their very essence as well as the pattern in the heavens, and in terms of man's place in that order as a co-worker with Nature – bears a profound message of hope; for in so far as man integrates himself positively and harmoniously in the Cosmic Order, he takes his active share in the building up or establishing on earth of the divine order as revealed to him in his deepest contemplation, and he becomes a 'godly' sharer in the divine work. That vision of the ancient sages has never been surpassed.

From the most abstract or highest to the most concrete or lowest, the sages embrace the universe as a totality, a wholeness, with the Absolute as the central point, the Source and Resolution of all

> whose shadow is death, whose shadow is immortality,[5]

beyond being, beyond non-being. Through a primordial contemplation activated by that supreme desire that a cosmos should manifest, through a sacrificial self-offering, emerges the active Deity, who in his turn sacrifices his oneness, his homogeneity, that the many should have an individual existence of their own; that this existence might reach its pinnacle in man, the offspring of, and the link between, heaven and earth. The projection from within without of that which is latent is itself an expression of the law of being, of the eternal rhythm which indwells and underlies all, the eternal ebb and flow.

The various stanzas scattered over the Vedic collections build up a vision of majestic grandeur. Only a few can be recorded here:

> Vast truth, mighty order, consecration, spiritual exertion, prayer, sacrificial offering, these uphold the earth (Athv. XII.1.1).

This stanza is of utmost importance in its summary not only of the Vedic vision, but of Vedic values. Our very earth is considered to be based, shaped and maintained on a 'holy foundation', that foundation of spiritual values that make for holiness.

Truth comes first, it is the very breath of that eternal Order which is mighty and terrible in the sense of awe-inspiring and unrelenting in its eternal rhythm, action and reaction, endlessness. To seek the truth, to live the truth, to be the truth, is to live in accordance with the *ṛta*, the Cosmic Order which is the truth of the Absolute, the truth laid out in the heavens, born of blazing *tapas*, the very first outbreathing of the Universe (X.190.1). Truth is the law of Deity. *Consecration*, that setting apart as sacred to the Deity, that dedication of the life to the higher ends of spiritual growth that leads to total submission to Deity, the first step on the path of the gods, or the path of no 'return';[6] *spiritual exertion* or *contemplation* (*tapas*), that straining of the whole being in one-pointed aspiration burning off the dross of mind and heart; *prayer*, the evocative power of the *word*, the soaring of the heart's innermost longing, sounding the hidden chord resounding through the whole being, thrilling through every fibre, opening the door to the infinite; and *sacrificial offering*, the ritual worship that sums up human endeavour and establishes communication between the deva and the human orders – these are all expressions of a spiritual venture that translates the divine order into human terms and falls under the category of 'holiness'. These are religious injunctions and fulfilments that spell out man's calling on earth; these are his truth that holds up the mirror to the Cosmic Order. The vision of holiness here embraces both the cosmos and man's dynamic role in it.

But the seers, being mystics, were perhaps even more preoccupied with giving voice to their vision of oneness, of the glories of the One, through the many, of the Whole, the Lofty, the Transcendent and man's relation to It, to which they gave many names and which they clad in many splendid forms:

> The fair-winged One, who is but one in nature, inspired seers, by their incantations, shape in many ways (X.114.5).

> To what is One the seers give many a name; they call it Agni, Yama, Matariṣvan (I.164.46cd).

I have known this mighty Celestial-Being (Puruṣa)
refulgent as the sun beyond the darkness;
only by knowing him does one overcome death,
no other way is there to go (Yjv.31.18).

Knowledge and what knowledge reveals are emphasised here: the refulgent Being at the heart of all and through whom alone death is conquered. For that Being pervades and transcends all:

Thousand-headed is Puruṣa, thousand-eyed, thousand-footed . . .
Puruṣa is this all, what has been and what will be and he is the lord of immortality which through matter transcends it all . . .
Such is his mightiness; but far greater than this is the Celestial Being. One fourth of him encompasses all beings, three fourths of him abide immortal in heaven.[7]

Transcendence and immanence of Deity, whether the latter be called Varuṇa, or Puruṣa, or Prajāpati, or even Aditi, are ever present in the mind of the bards. This is obvious from the Puruṣa sūkta quoted above, from verses addressed to Aditi, or to Varuṇa, or to Indra: Aditi is the all-containing, all-enveloping substance or Mother within whom evolve all things (Ṛgv. I.89.10). Varuṇa's surpassing splendour is beyond Nature's majesty:

Thy force and might and passion
neither these birds in their travelling can attain,
nor these waters ranging restlessly,
nor they who hedge in the hugeness of the winds.[8]

Although Nature is so to speak the mirror of the mightiness of Deity, the power, splendour and glory that man attributes to God and sees reflected in the vast panorama of sky, mountain and ocean, yet Nature does not equal Deity; Nature is but the garb with which Deity veils Itself. The same idea of transcendental might is carried over to Indra:

Vaster than days and nights, O Giver of increase,
Vaster than firmament and flood of ocean,

Vaster than bounds of earth and wind's extension,
Vaster than rivers and our lands, is Indra (X.89.11,
Griffith's trans.).

The might and majesty of the outer manifestations of Deity find their fullest expression in the seer's vision of their *oneness*, their ultimate fountain-head, where all are seen as links in the vast whole, emerging from and returning to the ONE:

The Seer beholds that Being hidden in mystery
In whom all find a single home; in That all this unites,
From that all issues forth. He, omnipresent,
Is warp and woof mid created things (Yjv. 32.8).[9]

Wholeness, completeness, flawlessness, inherent in the meaning of the word 'holy', are expressed in the Sanskrit word *sarvatāti*, the very boon asked of the gods.[10]

Furthermore, contemplation of *ṛta*, the transcendental Law, in all its ramifications, grants the vision of wholeness, a vision which the gods possess, since they are '*ṛta*-visioned', and which, through their activity, they actuate, i.e. manifest in the space-time world, e.g. in the balance of nature, the orderliness of the starry heavens, the eternal recurrence of all things; whilst men, catching glimpses of it, only try to realise it in their daily life, through their efforts at living in harmony with each other, at treating each other fairly, at striving towards truth, justice, righteousness:

Contemplation of *ṛta* annihilates transgressions.[11]

This verse is quite a revelation of the ṛṣi's knowledge of the implications of meditation.

A human being who is in harmony performs his duty perfectly, fulfills his obligations under the Law, lives in accordance with the inherent law of his being, is at perfect peace with himself and the world, is 'true', 'whole' and can commit no wrong. He is holy, (*ṛtavan*), he possesses *ṛta*. All converges on the fundamental issue of the working in harmony with the great Law or on its infringement. One results in well-being, one in suffering. Material prosperity considered as a result or reward of right living also points to this conclusion as underlying the whole Vedic conception of *ṛta* in so far as it is related to man:

> Sweet blows the breeze for the righteous man,
> sweet flow the ocean-deeps (I.90.6).

E. W. Hopkins summed up 'the ethical content of the Ṛgveda' thus: 'Morality is an expression of divine law; sin is opposition to that law. The sinner is one who is out of harmony with the higher spiritual environment, which encompasses and controls the world.'[12]

Two hymns, one in the Ṛgveda (X.191.2–4) and one in the Atharvaveda (III.30.1–3), express the vision of harmony as applied to humanity. Both emphasise concordance, togetherness, union. Justice among men can only be established when each man acts in harmony with his fellow man and with the whole of which he is an integral part. The Ṛgvedic hymn, as rendered by R. Panikkar, runs as follows:

> Gather together, converse, together. Your minds be of one accord,
> Just as in harmony the gods of old took their ritual share of oblation.
> United be your counsel, united your assembly, united your spirit and thoughts.
> A single plan do I lay before you; a single oblation do I offer!
> United your resolve, united your hearts, may your spirits be at one,
> that you may long together dwell in unity and concord.[13]

No more forceful call to peace among men, to togetherness, harmony, oneness, can be found among the world's religious literature.

Sinlessness is the prerequisite to human harmony, integrity is the hallmark of truth. It is its guarantee. So man asks the sun to 'declare us sinless', without blemish, i.e. whole, in the presence of Mitra, Varuṇa and other gods.[14] That Sūrya can do so is because the human heart is an open book to him and if he agrees to do so it is a proof that the man is in harmony with himself and all and thus worthy of communing with the Adityas.

The harmonious or holy man in following the law which is truth, *ṛta*, conserves the *ṛta* and is said to be 'true' (*satya*); hence the great power attributed to the declaration of truth. *Ṛta*, when it applies to the human moral order, is *truth*. He who follows *ṛta*, the path of righteousness, is a man of truth and has at his command the power that only truth can wield.

Truth, as we have seen, is the basis of Vedic ethics. A further aspect of this vast subject should be examined in this connection.

As formulated by E. W. Burlingame, the act of truth or truth utterance (*satyokti*) is 'a formal declaration of fact accompanied by a command or resolution or prayer that the purpose of the agent shall be accomplished'.[15]

How is it that such an act can produce the desired result as against any mere enunciation of fact? It is based upon one's complete fulfilment of one's being, function, occupation or duty where thought, word and deed are involved, hence absolute truth to one's self. As W. N. Brown beautifully puts it:

> The kind of Truth that is potent in the Act of Truth is not mere factual or existential or empiric truth, that is accuracy of statement concerning some fact or occurrence capable of comprehension by our senses and so capable of description in physical terms. Nor is it logical truth, truth reached by deduction from accepted premises. Rather, it is truth of life, truth in one's total conduct, truth in acceptance of responsibilities and performance of duty. It is truth as the metaphysical basis of the cosmic order, which for each human being and each deity is the sanction and ethical basis of his actions. It is the substance of his obligations or duty in the cosmos and it shares a quality which Polonius seems to have had in mind when he admonished Laertes 'To thine own self be true: and it must follow as the night the day thou canst not then be false to any man'.[16]

This does not mean 'that one can bend cosmic forces to his

will by performing his personal function with perfection',[17] as W. N. Brown states elsewhere. The purpose for which the act of truth is enunciated is in harmony with the Law; it cannot be otherwise for the truth of the human being is necessarily one with one or other aspect of his environment, otherwise it could not be true. Even the gambler's compulsion to gamble, as depicted in a Ṛgvedic hymn (X.34), must be expressive of certain forces at work around him which compel him to gamble, i.e. to stake, to take a chance, until such time as he has reached the depth of degradation and, unable to proceed any further, is left with nothing but 'his ten empty fingers' which he flings at Savitṛ, the Solar Deity, in a desperate gesture of truth, confessing his utter destitution. Because this confession touches an innermost chord of truth within himself, a realisation of his utter nakedness and shame, it works out his redemption; he is released from his compulsion and restored to sanity. The gods being true to law, whose statutes are truth, who are true through and through, who are '*ṛta*-visioned', whose laws are true (V.51.2), are therefore responsive to any human being who, through the truth in himself, appeals to the truth in themselves, and thereby comes within the circle of the great Truth, the Cosmic Order. Truth, *satya*, is the individual's – whether human or godly – expression of the Cosmic Truth, *ṛta*, which is the divine Harmony. *Satya* implies oneness with *sat*, essential being, one's life being the very expression of one's fundamental being, hence truth. It is here that we find that *satya*, truth, is included in *ṛta*, the Cosmic Order or Cosmic Truth, hence he who is true to his innermost being is one with the Cosmos, is 'integrated', 'whole', 'holy'. He can command the elements for he works in harmony with Nature. He is godly.[18]

It is thus man's task to translate the cosmic harmony as visioned by the seers into social order. Eternal law can be mirrored at the human level in justice, righteousness, hence strict morals, and integrity at the individual level. These form the backbone of the social order.

The healthy elements that go to make a strong and prosperous state are outlined in a Yajurvedic hymn in connection with the horse sacrifice:

> O Brahman, may there be born in this kingdom the wise priest famous for spiritual knowledge; may there be born the ruling man, heroic, skilful archer, piercing with shafts, mighty warrior; the milk cow, the draught ox, the swift steed; the accomplished woman (Yjv. 22.22).

Wealth and liberality should characterise the monarch and his duty is to protect his subjects and bring prosperity to them. The boast of King Aśvapati Kekaya as recorded in the Chāndogya Upaniṣad, whether it refers to an actual fact of his administration or not, may be taken as an indication of an ideal that was set before the ruler's mind:

> In my kingdom there is no thief, no miser, no drunkard, no man without a sacrificial fire, no ignorant person, no adulterer, and where is the adultress? (Chānd. Up. 5.11.5)

The ruler's prayer, as found in the Yajurveda, shows what were his duties and aspirations:

> Let my subjects be satisfied, my herds be satisfied, my people be satisfied, let not my people be needy! (Yjv. 6.31)

The *sabhā* was an assembly of the Elders where all problems were discussed and whose decisions could not be violated, membership of which was deemed an honour. A *sabheya* was a successful speaker in the *sabhā*, which implies verbal contests, skill and knowledge. The *samiti*, on the other hand, was the assembly of the people where representatives of the various classes could gather to discuss their problems. This is outlined in the Atharvaveda:

> May the Assembly and the Gathering, the two daughters of Prajāpati, concurrently aid me.
> May he, with whom I shall meet, co-operate with me; may I, O ye Fathers, speak agreeably to those assembled.
> We know thy name, O Assembly: Nariṣṭā verily is thy name. May all those that sit assembled in thee utter speech in harmony with me.

> Of them that are sitting together I take to myself the
> power and the understanding in this entire gathering,
> O Indra, render me successful (Athv. VII.12.1–3).[19]

The acquisition of wealth, whether by ruler or people, seems to be encouraged, but it must be done in the right way, not by devious means:

> Let a man think well of wealth and try to win it by the
> path of righteousness and by obeisance (X.31.2).

Honesty and the following of religious instructions are obviously deemed the best policy and will lead to well-being. For

> Indra seeks to help him who offers and gives,
> augmenting, not diminishing his possessions.
> Evermore increasing his treasure he places the pious in a
> realm inviolate.[20]

'Under the ṛta', as remarked by W. N. Brown, 'every human being too has a duty, a specific function called his *vrata*, a word which in the later language means "vow".'[21] Within the inclusive norms of *ṛta* there is much variety for human beings, the callings of men are numerous and some are somewhat humorously expressed in the following verse:

> Manifold are the callings of men, manifold their functions;
> the carpenter seeks what's broken,
> the physician the diseased,
> the priest the soma-presser . . .
> I am a bard, my dad's a physician,
> my mum throws the corn on the grind-stone;
> pursuing wealth with varied plans
> we follow our callings as the herdsman cows.[22]

Each occupation is an integral part of the social order and when duly performed contributes to the overall social harmony. We glimpse here not only the beginnings of the idea of *dharma* which was to be so forcefully expressed in the Bhagavad Gita, but also the divisions of labour of Vedic society, divisions which imply professional groups which in due time gave rise to classes (*varṇa*) and castes (*jāti*). That there were originally, in addition to such basic divisions as family (*kula*), village community (*grāma*) and settlement (*viś*),[23] three broad groups in the Vedic community may be inferred from certain hymns such as the following invocation to the Aśvins:

> Invigorate the spiritual-power (*brahman*) and the vision (*dhīḥ*) . . .
> invigorate the ruling power (*kṣatra*) and the heroes . . .
> invigorate the kine, invigorate the people (VIII.35.16–18).[24]

We find here a threefold division of society expressed in the pursuit of knowledge yielding spiritual leadership, power over thought, education, culture, hence the priesthood; the wielding of temporal power and protection of the social order, hence the ruling class or government; the various trades of the common people, the economic aspect symbolised in the cattle, hence the productive class. This threefold division of society – priests, rulers, folks – appears incidentally in various addresses to the gods, such as the following:

> I call upon ye, Indra, Ṛbhus' lord . . . Varuṇa, Mitra, both ye Aśvins, let them foster for us wealth, devotion, victory[25]

wealth meaning well-being, which with 'healing' is the gift of the Aśvins, devotion or vision leading to song, pertaining to the seers and therefore the priestly class, and victory in contests to the warrior class of whom Indra is the leader.

The divine prototypes of the ruling class (*kṣatriyas*), the gods that enforce the keeping of the law and thereby protect the ordinances and well-being of human society, are Mitra and Varuṇa: they are lords of the luminous dominion of lofty *ṛta*.[26] Indra, the *kṣatriya par excellence* whose task is to fight all opposition to Order is however called *brahman*, and the 'forwarder of prayer' (VI.45.7), the meditating god, thereby showing that both functions were considered closely knit.

The four classes which came to be distinctive of Hindu society are enumerated in the *puruṣa sūkta* (Ṛgv. X.90.12), considered one of the later compositions of the Ṛgveda. Issuing from various parts of Puruṣa's body, they form the limbs of the divine Being, differing in their functions, each necessary for the welfare of the whole; one in their divine origin and one in their common purpose: manifested harmony. It is here that we meet the word '*sūdra*' for the first time, 'worker', or 'tiller of the soil', used but once in the Ṛgveda whereas it appears regularly in the Atharvaveda, perhaps because the latter is the Veda of the people rather than of the priests as the former is, or perhaps because the tillers as a class are by now more sharply differentiated from the bulk of the ordinary people, or trades (*viś, vaiśya*). However, one other verse in the Ṛgveda shows the distinct separation of society into four classes:

> One to high sway, one to exalted glory, one to pursue his gain, and one his labour;
> All to regard their different vocations, all moving creatures hath the Dawn awakened (I.113.6).

That the classes were accepted as a fact of nature may be gathered from the several prayers of the Atharvaveda and the Yajurveda, such as the following:

> Make me . . . dear to brahman, and kṣatriya,
> both to śūdra and to Aryan (Athv. XIX.32.8).

> Give brightness to our priestly folks
> make bright our ruling folks, bright our traders and tillers,
> and through that brightness give brightness to me (Yjv. 18.48).

These are prayers for the collective welfare of society of which many are extant in the Yajurveda and the Atharvaveda. In the former, the bard, addressing the gods to come in harmony to grant him his boon, goes on:

> May the seven assemblies and the eighth which unites all beings fill our paths with what we desire . . .
> So that I may speak the blissful word to the masses of the people,
> to the brāhmaṇa, and the kṣatriya, to the śūdra and the vaiśya,
> to our own men and to the stranger (Yjv. 26.1,2).

The rigidity with which in later times each of the four classes was differentiated from the other does not seem to have been prevalent in the early Vedic age. In a prayer to Indra, the bard expresses his wish to combine the three main social functions in himself, and indeed to rise to that of a seer:

> Wilt thou not make me a herdsman of people,
> wilt thou not make me their ruler, O liberal Soma-drinker,
> a seer, drinker of Soma juice, a skilled-controller of
> immortal riches? (III.43.5)

A song of the Ṛgveda (IV.57) dedicated to the lord of the field shows an ideal world of simple happiness where labour becomes a constant communion with the divine powers that pervade all, as the worker dedicates in prayer every move he makes, every tool he uses, every spot he ploughs. The whole song is a consecration of the labourer and the soil.

> In contentment may men and oxen both plough,
> in contentment the plough cleave the furrow,
> in contentment the yoke be securely attached
> and the ploughman urge on his oxen!
> Ploughshare and plough, to our chant be propitious!
> take of the milk you have made in heaven
> and let it fall here on this earth!
> Auspicious furrow, we venerate you.
> We pray you, come near us to prosper and bless
> and bring us abundant harvests.
> May Indra draw the furrow, may Pūṣan
> guide well its course! May she yield us milk
> in each succeeding year![27]

The ideal of harmony, the mirror of the vast Cosmic Order, is found emphasised in all the Vedas. The ideal of family concord which is the root of social harmony is given full expression in a song of the Atharvaveda:

> The union of hearts and minds and freedom from hate
> I'll bring you.

> Delight in one another as the cow in the calf that she has borne.
>
> Let son be loyal to father, and of one mind with his mother;
> let wife speak to husband words that are honey-sweet and gentle. . . .
>
> I will make the prayer for that concord among men at home
> by which devas do not separate nor even hate one another. . . .
>
> Common be your water-store, common your share of food;
> I bind you together to a common yoke.
> United, gather round the sacrificial fire
> Like spokes around the nave of a wheel.
>
> With your common desire I'll make you all
> have one aim, be of one mind, following one leader,
> like devas who preserve their immortality.
> Morn and eve may there be a loving heart in you.[28]

Righteousness in man, harmony in his community, such is the basis of the ideal social order as envisaged by the sages so that they could say

> Where spiritual and ruling powers move together in unison, that world shall I know as holy where dwell the shining ones together with the Flame (Yjv. 20.25).

However far away from this order man might stray, he could not fall completely beyond the fold, for the law, as administered by the gods, encompasses all in its mighty sweep and brings back the sinner sooner or later into the fold:

> Beyond the statutes of the devas no one,
> even if he had a hundred souls, can live (X.33.9ab).

Hence the great hymn to peace in the Atharvaveda (XIX.9) and the well known lines in the Yajurveda:

To the heavens be peace, to the sky and the earth,
to the waters be peace, to plants and all trees,
to the gods be peace, to Brahman be peace,
to all men be peace, again and again
peace also to me! (Yjv. 36.17)

O earthen vessel, strengthen me.
May all beings regard me with friendly eyes!
May I look upon all creatures with friendly eyes!
With a friend's eye may we regard each other! (Yjv. 36.18)

Man's confrontation with himself

This third level of the Vedic vision of *ṛta* shows us puny man confronting himself: his possibilities and his limitations, his aspirations and visions, his inadequacy and complacency; his humility and pride, his hopes and despair; his fears and joys, his suffering and exultation; his self-betrayal and doubts, his revolt and his submission; his realisation of his inherent freedom and power to know and his claim to his divine birthright, to choose freedom; in a word, his role in the vast scheme of the universe in which he finds himself seemingly caught up for better or worse as though in the cogs of an ever revolving wheel. And puny man, through this very facing, finding and accepting of himself, comes out magnified.

Ṛgvedic man probed the human condition, faced its dark and its light side, accepted it as part of the course of all things, and firmly placed man as an inherent constituent of the cosmic order which he, as well as the gods, helps to further. The struggles, the pitfalls, the progress and the drawbacks, are all essential steps in the vast field of experience offered by the cosmic order, the means being conflict, the friction of opposite polarities, the end being complete harmony of all the integral parts. At this juncture man is the epitome of the darker side of the cosmic order for he stands at that stage where conflict prevails prior to that further stage which the deva commands, where harmony has been achieved and the tension between the polarities resolved. In this respect it is interesting to note that Indra found for man the way to 'fulfilment', *iṣṭaye*, through conflict, *yudhā*, (Ṛgv. X.49.9). But man has it in himself to become the epitome of the lighter side of the cosmic order, by integrating himself in the harmony, hence uniting in himself the whole and thereby being the perfect mirror of that whole which he contains potentially in himself.

The search for himself, both outward and inward, was obviously part of Vedic man's vision. The keynote of this search and its result, the bursting of man from his bonds and the call to freedom, is well expressed in a hymn from the Atharvaveda, as rendered by Dr Panikkar:

> Open yourself, create free space
> release the bound one from his bonds!
> Like a new born child, freed from the womb,
> be free to move on every path (Athv. VI.121.4).

Part IV

ṚTA AS THE SACRIFICE

TWELVE

The meaning of the Vedic sacrifice

Introduction

As *ṛta* designates the cosmic order in its metaphysical and physical sense and that order as reflected in human society in the accepted ethical norms of human behaviour, so also in religious practices it refers not only to the harmonious sequence of ritualistic function, but to its inner meaning: a dramatic, miniature and symbolic representation of the cosmic process – that process according to which all parts work in harmony with the whole towards a common end, the end being the establishment in our space-time dimension of the celestial *ṛta;* that process which may be summed up in the law of give and take which represents life and which is epitomised in the circuit of birth, growth, decay. The child takes in order to grow; the adult gives or returns in his own individual way what he has taken and assimilated. That which is given is in one sense a deprivation, a sacrifice, yet it is a restoration creating a balance, for there can be no giving without a previous receiving or taking: the rite of sacrifice in its inner structure is *ṛta,* the eternal balance, the universal order, the divine harmony.

The ritual as a mirror of the cosmic process, hence the sacrifice as man's expression of law and order, is of paramount importance in the Ṛgveda. The whole sacrificial rite was conceived as a bringing together of gods and men so as to enact dramatically the cosmic order. As stated by A. C. Bose:

> Rita is *rite* as well as *right*. As a ritual Ṛta signifies the orderly performance of the ceremonial part of the worship which is a complicated form of action, and therefore possesses the attributes of art. So like the aesthetic form, the form of the Yajña also conforms to laws of order. And the order observed at the ritual is a symbol of Eternal Order, including the eternal statutes of moral life.[1]

When we approach the Vedic sacrifice we have to discard current notions of sacrifice, including the idea of blood sacrifice, and face a completely different conception which at first sight may seem difficult to grasp. So the meaning of sacrifice, its vital importance in the Ṛgvedic religious cult, and its identification with the cosmic order will now be examined.

The primordial sacrifice as the genesis of all

Sacrifice, like *ṛta,* is intimately connected with the beginnings. In this respect, R. Panikkar's dichotomy of the trend in human intuition concerning the origin of all into two possibilities explains the process of manifestation, as viewed by the ṛṣis, along two lines, in connection with the verb *to be* which comprises the twofold notion of the state of *being* from which all becomes, and the notion of *becoming* which leads to being. The first concerns

1 *be* as the 'infinite reservoir of possibilities', infinite Being, the BE from which all becomes, unfolds, develops from what is already there.

The second posits

2 *sacrifice* as '*Becoming* is a coming into *Be,* into being'. 'In the beginning "was" Sacrifice. It is the Prajāpati-sacrifice, in mythical terms, which gives birth to Being.'[2]

There is here a profound intuition of the original significance of 'sacrifice'. The mind has to try to stretch beyond being to imagine a state out of which being emerges through sacrifice, through the limitation of that original homogeneous state, the latter being left undescribed, beyond the mind's highest flight. This dichotomy in the viewing of the origin, however seemingly contradictory, does not appear to have been held so separate in the vision of the Vedic seers. Thus the Absolute of the *nāsadīya* hymn is beyond Being and Non-being, yet out of It arises some power which, through the very action of *tapas* which itself is a sacrificial, self-limiting, contemplative act, generates Being. Both *Be* and *Becoming* are two aspects of the one reality related through *tapas*.

In the famous hymn of the primeval sacrifice of Celestial Man – Archetypal Man in his essential divinity – called *puruṣa sūkta*, the divine essence is sacrificed that the many may be. So Dr Panikkar goes on: 'The texture of the universe is sacrifice, which is the act *par excellence* which produces all that is.'[3] Sacrifice, the limitation for expansion in another dimension, is the divine 'action' *karman*, the original Vedic meaning of the word, an action which man imitates in the sacrificial ritual. In the *puruṣa sūkta* the gods are shown as performing a sacrificial rite by immolating Puruṣa, Archetypal Man, as a result of which the world as we know it comes into being. Accordingly, the primeval spirit, instead of remaining one, homogeneous whole, was fragmented into the many forms pervaded by the one life, its own life-breath. In the mythological language of the Ṛgveda:

> Three-fourths of the Heavenly man ascended on high,
> one-fourth remained here below. Thence he pervaded
> all things, animate and inanimate.
>
> The spring was the melted butter, the summer the fuel,
> the autumn the oblation (X.90.4 and 6).

The various castes and the various planetary bodies are born from the various parts of Puruṣa's body.

As stated rightly by Eggeling in his introduction to volume 43 of *The Sacred Books of the East*:

> In this primeval – rather timeless, because ever-proceeding – sacrifice, Time itself, in the shape of its unit, the Year, is made to take its part, inasmuch as the three seasons, spring, summer and autumn, of which it consists, constitute the ghee, the offering-fuel and the oblation respectively. . . . Prajāpati who . . . [in the Brāhmaṇas] takes the place of the Purusha, the world-man, or all-embracing Personality, is offered up anew in every sacrifice; and inasmuch as the very dismemberment of the Lord of creatures, which took place at that archetypal sacrifice, was in itself the creation of the universe, so every sacrifice is also a repetition of that first creative act. Thus the periodical sacrifice is nothing else than a microcosmic representation of the ever-proceeding destruction and renewal of all cosmic life and matter.[4]

For R. Panikkar,

> The sacrifice of the Cosmic Man signifies divine transcendence investing humanity. This universal sacrifice possesses a twofold dynamism, for it includes a sacramental downward movement of the All toward the earth and a sacrificial upward movement of the world toward the All; these two aspects are inseparable one from the other precisely because of the unity of the integral sacrifice.[5]

The world order, its regular movement, its activities converging on the one aim of fostering life, all this, to the ṛṣis' mind, was laid down in heaven from the very beginning of manifestation:

> By that sacrifice the gods paid homage to the sacrificial-host. Such were the first ordinances (X.90.16ab).

Of this mighty oblation the human ritual is the microcosmic reflection. The underlying factor of the interlinkedness of all participants, their joining forces in a common re-enactment of the universal process of creation – the sacred 'work' (*apas*) *par excellence* – is a peculiar characteristic of the Vedic conception of the sacrifice, a characteristic which explains its intimate link with *ṛta*, but one completely missed by nineteenth and early twentieth century Western exegesis.[6] The human rite is a tribute paid to the gods and through them to the divine life of which these are the ministers – that life, the Puruṣa, through and by means of which all creatures live.

The original idea, as implied in the *puruṣa sūkta*, simply expressed the offering, by the gods, of the divine life, the *puruṣa* as essence of all, the life-blood for the many; i.e. that divine life which courses through all and animates all. It is split up in many forms but one in essence.[7] The giving and the taking, that universal rhythm which is part of the vast becoming, *ṛta*, is present in this primeval *yajña*, whence derive all the others. It is the supreme offering of self which is itself an act of creation, and a making *sacred*. M. Bloomfield rightly affirms that 'The sacrifice is the dominant note of Vedic life, as far as it is revealed in these ancient documents.'[8]

The eternal circuit of give and take, of assembling and separating, of uniting and dividing, of diverging and converging, the basis of manifestation, the constant renewal of life through self-offering, i.e. through death, imaged as the sacrifice of Puruṣa in the Ṛgveda, the divine life immolated at the foundation of all things, is equally in the Brāhmaṇas represented as the dismemberment of Prajāpati, the equivalent of Puruṣa, the Lord of beings who is recomposed by the gods.

> Inasmuch as he (Prajāpati) created the gods, thereby he is the father of the gods; and inasmuch as the gods restored him, thereby the gods are his father.[9]

There is neither waste nor death in reality in so far as the essence of the one life energy is concerned. Nature shows us the truth: water rises as vapour from the oceans and the rivers under the action of the sun; this in due time becomes clouds that drop back on earth as rain which fertilises the soil and fills the rivers that in turn fill the oceans; and so the circling process goes on for ever. That which is taken up in one form is returned in another; the eternal flow of life seems to enter into dissolution; through the disappearance of a particular form it reappears into another but to disappear once again (cf. St Paul, the seed cannot bear fruit except it die).

So the Śatapatha Brāhmaṇa tells us:

> For when the gods in the beginning seized (slew) a victim, then as it was drawn upwards its sacrificial essence flowed downwards, and from it sprang the tree (3.8.2.17).

This is the *tree of life*, the *sacrificial post*, *Puruṣa* himself.

If, with W. N. Brown,[10] we realise that Puruṣa is not merely Archetypal Man, or Cosmic Man as a 'primitive' image of the world, or the 'male' as the accepted interpretation has it, but also the 'vital element' or 'soul' of plants (*puruṣa auṣadhinām*, X.51.8) which is Agni, as of all creatures, of the waters, and indeed of the world, then the link between Puruṣa and Agni who is all this as well as the sacrificer who sacrifices himself to himself[11] and the Lord of immortality, becomes perfectly evident.

The 'eater eating is eaten'[12] is a truth of nature, 'pregnant and terrible' to use Śri Aurobindo's expression, which the ṛṣis observed, and which we find emphasised again and again in their scriptures, however crude this truth may seem to our sophisticated minds. Its deepest meaning is the sacrifice.

Sacrifice is a dying and a becoming; this is creation; without dying there can be no becoming, no renewal, therefore Prajāpati sacrificed himself to himself that the world might be. As H. Aguilar rightly remarks, 'One fails to animate what one has created unless transmitting to it one's own life.' This implies 'the fundamental capacity of everything to immolate itself, to sacrifice its own separateness in order to regain the original unity or . . . in order to reconstruct the body of Prajāpati.'[13]

The basic significance of the sacrifice

The key to the basic significance of the Vedic sacrifice may be found in the words *tapas*, *apas*, *karman* and *yajña*. The Vedic sacrifice is a work (*apas*); an action (*karman*); a participation, communication-in-worship, communion, these being expressed in the one word *yajña*. These three basic concepts can be examined in relation to *tapas*.

We have seen how, in the Vedic vision of the beginning, *tapas* was the supreme means to the supreme end of creation, as a result of which truth (*satya*) and order (*ṛta*) manifested. The divine creative act was performed as the arousal of that flame-power or all-burning energy, *tapas*, caused desire to develop the seed of consciousness and project outwardly that which was held inwardly, *in potentia*, hence intelligent activity, hence the cosmic order:

> That One emerged through the power of *tapas* (X.129.3d).

This original divine act, this supreme incandescence whereby the universe was kindled to existence, set the seal upon every subsequent act of creation and endowed each act with some kind of sacredness which no amount of repetition could remove. All things are achieved through *tapas*, says the Atharvaveda:

> Through the tapas of self-control the king protects his kingdom. Through the tapas of self-control the teacher attracts his pupils. Through the tapas of self-control the devas drove away death from themselves.[14]

As earlier noted, *tapas* refers to the highest contemplative exertion at the creative, godly level; the thought of expiation, mortification, self-torture, does not enter into its highest conception. Only at the human level is there a need for that kind of purification which may be thought of in terms of mortification, for the human, having both the corrupt and the incorrupt, has to purify himself of the former if he would know the latter fully. But even at the human level mortification and self-torture are extreme activities which *tapas* in its essential meaning – the arousing of the flame divine – did not originally imply. *Tapas*, at the human level, as contemplative exertion, that intense spiritual focusing which occurs in deep meditation, brings about inner incandescence as a result of self-imposed purification and one-pointed concentration of the mind; this implies limitation in one way or another for further expansion. This is sacrifice in its deepest sense of making sacred, an action that is both self-limiting and self-expanding, an offering of self to receive or contact a greater Self. The very basis of creation is thus sacrifice, i.e. self-imposed limitation, ordered activity, self-offering, not only at the primary stage, the original act of creation through *tapas*, but also at the secondary or godly stage and at the third or human stage.

Through the practice of *tapas*, the ṛṣis are able to 'see', i.e. they received their vision, inspiration, revelation; this means that they kindled their inner fire, hence are said to be born of *tapas* (*tapojan*) and 'skilled in a thousand ways' (X.154.5). This is not just a vague reference to inspiration, but a specific inner exertion when all the powers of the human being, the senses, mind and emotions, are bent upon one inner focus and become integrated like 'the rays of the sun' focused in a 'lens', hence kindled to their utmost possibility. Mention of *tapas* as religious fervour, the inward turning of consciousness brooding upon itself, or inner incandescence, thereby arousing the secret creative Fire 'whom men find in their heart when they have sung their mantras which they fashion there' (I.67.2), abounds in the Ṛgveda, but was sorely misunderstood until only recently.[15]

Thus not only are the seven primeval ṛṣis described as practising *tapas* (X.109.4b), but the patriarchs, the Angirasas, create through *tapas* (X.169.2c) and Indra wins heaven's light through it (X.167.1d). The gods gained immortality both through *tapas* and through sacrifice,[16] the former being an expression of the latter.

There is frequent offering on the part of the priest of his 'toil' or 'sweat' as the token of the sacrificial work or his own devotional attitude. This is surely more than the mere exertion of kindling the fire. In the service of Agni the bards offer up their 'sweat' (V.7.5), 'heating their head' (IV.2.6) as they brood in contemplation and set themselves on fire. This verse was rendered thus by C. J. Blair:

He who carries fuel for you, sweating, or who heats his

> head because of (devotion to) you, do you Agni be his self-strong protector; protect him from all evil.

To which C. J. Blair adds the following commentary:

> Heating the head as a devotional activity is here on a par with carrying fuel and sweating. The same activities which caused sweating must have given a sensation of heat to the whole body, but, because the head was the center of devotional thought, the heat which was felt there was given a special value and significance. In other words, a man's head was hot because of particularly energetic devotional thought.

As is well known the heart was considered the centre of devotional thought. The same author continues:

> If Soma's effect was intoxicating and made the drinker think himself a king, or a Ṛṣi, etc., the fire's effect might have been to overheat the head enough to create hallucinations which the ritual priest took for wisdom and insight. This is Hauer's opinion (Die Anfänge der Yogapraxis, p. 111), but I have no reason to believe that head-heating meant anything more than mentally energizing for the purpose of carrying out the offerings, and rituals, and singing the hymns in the proper and correct way.[17]

Heating the head is a quaint rendering of that inner brooding which is meditation and which results in kindling that most secret inner fire which later literature was to call the serpent fire. When that fire is aroused through the practice of meditation and reaches the brow it opens up 'vision', 'insight', hence 'illumination', the feeling of 'immortality', complete 'self-control' and 'command' over all. To reduce this to mere hallucination evidences complete misunderstanding of the whole process and indeed of what yoga means. The ṛṣis' vision of cosmic order evidences profound insight which cannot be born of hallucination, their social and ethical doctrine profound wisdom, all of which is hardly fathomed by our shallow-minded generation.

In the later literature the 'inner sacrifice' is mentioned as that offered to the gods. In the Kauṣītaki Brāhmaṇa Upaniṣad 'breath' is said to be sacrificed to speech and speech to breath:

> these two are unending immortal oblations; whether waking or sleeping one is sacrificing continuously (II.5).

This is obviously a reference to the practice of *prāṇāyāma*.

So the sacrificial rite involved a making sacred through purification, an offering of the essence through burning off the dross; both at the personal and ritual level it implied a self-discipline through applied orderliness, the making of a cosmos out of chaos. This making sacred, this sacrifice, to the Vedic mind, was essentially 'creation'; it was *action*, as the word *karman* used originally for ritual connotes, it was *apas*, the sacred work.

The main word used for sacrifice is *yajña* which is derived from √*yaj* meaning 'to worship', 'to pay honour to'. It is an act of relatedness linking one level to another, making use of the concentration of mental power which 'yokes'[18] the mind to the object of worship. In the sacrifice the mind yokes itself to a vision of the beyond, or to one god or another, or to the conception of the cosmic order, hence the interrelatedness of all, the rite being a framework, the means whereby the mind is directed towards and, possibly by means of it, penetrates into the transcendental realm. As expressed by R. Panikkar

> Worship does not consist solely in prayer . . . it is action, an action by which duality is transcended and dissimilarity banished. This act contains within itself, essentially, a sacrificial aspect, a death and a becoming, a doing, *karman*.[19]

This sacrificial action which is a death and a becoming is the creative act *par excellence*. Ritualised in the sacrifice, it is a sacred work performed according to certain well-attested rules, hence its identification with *ṛta*. It should therefore not be taken in the sense of the shedding of an innocent victim's blood to propitiate a deity. Indeed it is the current meaning of sacrifice as self-deprivation, or animal immolation, or vicarious atonement, that has militated against the understanding of the Vedic sacrifice.

The Ṛgvedic seers conceived the sacrifice as a universal process to which all things are subservient, an eternal give and take. This, at the lowest level, manifests as what appears as a constant 'preying'; at the highest as a constant 'self-gift'. Hence the universal application of the Ṛgvedic riddles:

> This sacrifice, the hub of the universe (I.164.35b)

> O Agni, sacrifice with thy mouth to thine own self (VI.11.2d)

i.e. offer up through thy mouth (through the action of receiving and burning) that which will come back to thyself, beyond all aspects or forms, for all is one. Here the circuit of giving, taking and returning the gift, has gone full round and recalls the idea of the descent of the spiritual life into the limitation of the material and its reascent enriched by experience.[20]

> O Self-sustaining One, offer up thyself to thyself for exaltation (X.81.5cd).

Of this grand cosmic conception Agni is the divine ministrant, and through Agni, the 'immortal guest in mortal houses', man partakes in the eternal sacrifice. In this dramatic enactment there is built up a relationship between terrestrial man and those divine powers he invokes and seeks to serve, there is started an interaction that mirrors the interplay between the two main poles of manifestation, spirit-matter, earth-heaven, positive-negative, out-going-in-going, father-mother, the eternal rhythm of the universal order.

The ritual as a means of communication, a coming together for a sacred purpose

The Vedic ritual was a communion with the higher powers mainly through the agency of Agni as the high priest or oblation bearer. A whole hymn of the Ṛgveda (III.27) bears this out. Verse six states that the officiants

> worshipping with true-vision-inspiration have prepared Agni for our aid.[21]

i.e. have kindled the sacrificial fire to perform the ritual for human help.

The rite is often considered a 'ship' or a 'chariot'; it is a means of communication, of bringing closer the two shores, that of the hither or terrestrial realm, and that of the beyond, or godly realm.[22] The worshippers' songs and homage, their visionary thoughts translated into hymns, their praise and prayer (*brahma*) frame as it were a 'vessel' that can be filled by the divine influence, blessing, riches:

> Through these our flashing-songs do thou turn towards us like heaven's radiance.[23]

The action is reciprocal and is summed up in the sacrificial chariot travelling hither and thither like

> conquering chariots gaining wealth, giving aid.[24]

The emphasis lies on blessings poured out through the power of visioning and offerings sent to the gods through that same power:

> Vision was established as a help towards the seeking of riches.[25]

So Agni is enjoined to 'shine riches on us through our fine-praise-songs' (V.25.3d). The vision that is a means of revelation acts as a focus for entering into a different dimension and thence securing blessings, knowledge, and so on, the riches of the gods.

The ritual is also compared to a web (*tantu*) woven by a hundred and one ministers (X.130.1). It is a spinning; the spinning of a web symbolic of the fabric of the universe which man spins from the sacrificial enclosure right to the vault of heaven.[26] It is an 'ancient' participation. Thus Agni is asked to spin out the 'ancient thread' (I.142.1) connecting heaven and earth. This is the bond that binds man, the earthly being, to heaven. At the level of ritualistic religion man shares in the universal work in so far as he re-enacts the primordial sacrifice sacramentally as symbolic of the cosmic life. A colourful summary of the whole idea – as translated by J. Gonda – is given in the following stanza:

> The warp of ṛta has been extended in the sieve on the tip of the tongue by Varuṇa's *māyā*. Only the 'wise ones' who desired to attain that have reached it.[27]

The Cosmic order is spun in the very sieve where the preparations are made for the ritual, as well as on the tongue where the words are pronounced, all due to Varuṇa's magic.[28] J. Gonda offers this explanation: only those 'hymns [which] have successfully passed through the sieve because they are in harmony with the reality underlying the phenomena do not infringe the laws and rules obtaining in the universe.'[29] This could imply that a communion has been established and therefore that a penetration of human consciousness into another dimension of being has become possible. Only the wise ones have managed to do so.

The ritual establishes such a close relationship between man and god as brings the human into direct contact with his god. The former claims his entrance into the divine domains and when this is refused, as seems to have been the case with Vasiṣṭha, the bard demands the reason for such estrangement:

> And now what has become of our friendship of old, when we two kept unhindered company? O Varuṇa, law-abiding, I entered thy mighty mansion, thy thousand-gated home (VII.88.5).

The ancient sacrifice does afford us an insight into a grand conception of the universe where the interdependence of all things is recognised and man viewed as an active participator, a truth left completely unobserved by nineteenth century exegesis. The lack of understanding of this aspect on the part of nineteenth century scholars led them to a series of erroneous assertions, as for example this one, that 'the same presumptuous assumption that the gods depend on earthly sacrifice is often made'[30] as in the period of the Brāhmaṇas literature. The 'dependence' was rather the 'interdependence' of all forces, an interaction of all those factors that play a vital role in the order of the universe. If later on the conception of 'interaction' was changed to mere 'dependence' of the gods on human sacrifice and the sacrifice degenerated into an exchange of gifts, in an attempt to make the priesthood all powerful, such is not the original idea as reflected in the Ṛgveda.

We have to come to the mid-twentieth century to find a much better understanding and assessment of the meaning of the Vedic sacrifice.[31] J. Gonda considers that the whole idea evidences

> one grandiose theory of human welfare and salvation advanced in order to teach man how to bring, by ritual means, the microcosm of human existence into harmony with the powers governing the universe . . . so as to gain welfare, health, a full term of life.[32]

We are at last coming closer to the realisation that the ṛṣis had recognised, through their vision, the interconnectedness of all things and had summed this up in the vision of the cosmic order, *ṛta*, which they re-enacted in the ritual. In this respect R. N. Dandekar gives a fine summary of the Vedic idea of sacrifice-cosmic order:

> A sacrifice, it was believed, was not a mere propitiatory rite but it possessed a profound cosmic significance. A cultic act established a magical rapport with the entire cosmos. A sacrifice was not merely a representation in miniature of the cosmic order, ṛta, but it was also a necessary condition for the proper working of the cosmic order. The performance of specific cultic rites was made to correspond with the rhythmic course of nature.[33]

As asserted by H. Aguilar, 'The cult is the natural milieu for the Vedic theophanies in keeping with the essentially priestly character that the gods exhibit in the Śruti.'[34] So when

> companion Brahmans assemble together to worship with flashes of insight shaped in the heart (X.71.8a),

these are said to leave far behind them those that speak without real knowledge. This points to a coming together for a pure and sacred function, a worshipping together in the heart where the divine voice is heard. In such moments the poets tracked down the sacred word (*vāc*) right to the very core of their heart and expressed her in various outbursts of poetry (X.71.3).

A. B. Keith in his survey of Vedic sacrifice complains that

> the theory of the sacrifice as a communion with the deity, whether by direct rite of eating with him, or as a sacrament through eating a victim . . . is not recognised as such in the formulae of the ritual; we simply find nothing of the view that the worshippers are eating together with the god in order to renew their relationship.[35]

This misses the whole point of Vedic sacrifice. If there is no emphasis on the 'eating together' but rather on the 'offering' together through 'mind and heart' by way of the Soma outpouring, the oblation, and the chanting and visioning, it shows the Vedic sacrifice as a communion at the level of the heart and mind rather than the mere physical body, and takes it far beyond any kind of 'primitive sacrifice'.[36] So,

> along the path of ṛta, from the sacred work, speed forth the inspirations to Indra-Agni (III.12.7).

According to the traditional order which has become the path of *ṛta,* these inspired revelations are offered up to the gods. By such means the poet will be able to enter into communication with the devas. This is the sacred work (*apas*). A. B. Keith goes on:

> the conception of the sacrifice as a communion of men both *inter se* and with the god and as a sacrament through the feeding on a victim which is really an embodiment of the god, were present only in germ, and they do not seem to have generated the consciousness of the sacramental nature of the offering

evidencing a complete failure to grasp the essential nature of the Vedic sacrifice. The sacredness of the rite is made plain in the following:

> Upon our knees let us approach thee with adoration as thou shinest forth in thine own domain,[37]

Agni's own domain being the altar where the ritual is performed. Agni, the oblation-bearer to the gods, is 'served with homage and oblation':

> With our homage rich in oblation, let us arouse that performer of the sacrifice by the path of ṛta.[38]

> Reverencing with oblation in our spirit Agni-common-to-all-men, the truth-bearer, the light-finder.[39]

> All indeed partake of thy divinity as they keep each in his own proper way, the eternal law.[40]

The eternal law (*ṛtamamṛtam*) – the cosmic order, the eternal circuit, the law of give and take, of receiving and returning, of exchanging, of balance – this partaking of the god's divinity occurs only in as much as the divine law is kept. The stress is on the partaking of divinity by all in the very keeping of the *ṛta* both as the eternal order and the eternal rite, for *ṛta* here bears both meanings.[41]

The poet who receives his inspiration from the gods, offers it back to the gods:

> From the seat of *ṛta* I send thee a mind-yoked inspiration (VIII.13.26b).

In a similar vein, to the partaking of Soma in the sacrifice is attributed the stimulation of spiritual insight, hence revelation:

> Wisely have I taken my share of that honeyed oblation generous in spiritual insight (VIII.48.1ab).

Partaking of divinity, receiving of spiritual insights, illumination, are effects of the coming together in the sacred function of the ritual.

When Agni 'the wise son of heaven', the 'redeemer' (*kilbiṣaspṛt*) is specifically asked to

> 'bring the gods hither one by one for worship' (III.25.1d)

the implication is that communion between humans and deities is to be brought about through the agency of Agni, hence through the ritual.

Heaven and Earth themselves are asked to accomplish their task (√*prakṛ*) through the poet's new songs, 'in the matrix of *ṛta*'.[42] At the cosmic level, as previously shown, the parents are thought of in terms of first engendering the gods and then as keeping within their norms all creatures as these travel between the Father and Mother (X.88.15). Between these two also journeys the sun in accordance with the rhythm of the cosmos, i.e. established law (I.160.1d). At the ritual level – the dramatised expression of the *ṛta* – they are asked to help in the gathering together of both gods and humans, hence forward a harmonious togetherness. They

> sit around the sacrifice and promote the *ṛta* of Mitra (IV.56.7).

We glimpse in this togetherness of gods and men a social and spiritual purpose furthered by the rite, both in the assembling of men to uplift their heart and mind in a joint effort, aspiration and communication, and in the calling upon the gods to participate in, promote and protect men's endeavours.[43] Thus is the great *ṛta* kept going.

> Serving the *ṛta* with the *ṛta* (V.68.4a)

also alludes to the means whereby the *ṛta* in its cosmic sense is best expressed by the *ṛta* not only in its human ethical sense but also in its ritualistic bearing; the means whereby Mitra and Varuṇa have attained their might being their conforming to the laws of the universal order as a whole and to those laws as particularised in their symbol, in the god's case, the cosmic sacrifice. Thereby these gods

> have attained the lofty throne (V.68.5c).

Furthermore, three of the main elements of the rite, the

word, or holy syllable, the *fire* which burns up the dross, and the *purified oblation* borne by the fire to the gods, are said to be 'god-given' and thereby stamped as sacred:

> First the invocation, thence the flame,
> thence the oblation, was begotten of the gods (X.88.8ab).

Of particular interest is the order of precedence: the recitation of the hymn (*sūktavākam*), i.e. the poetic utterance which springs from the depths and is directly inspired by the gods; the fire both purifying and energising, destroying and creating out of the ashes of what is eliminated (perhaps here also specifically the inner fire kindled by inspiration-aspiration – poets are fired on to greater and greater eloquence); finally the offering which consists of that very inspiration purified by fire given as an oblation, together with materials such as ghee, milk, soma, to those same gods that caused the original inspiration. The best offer seems to be a song, a hymn, a prayer, from the worshipper's heart, but all are expressions of a god-given vision now returned to the god fashioned by the poet, made the poet's own and fire-proved. A verse from a eulogy to the gods (III.55.8c) makes this quite clear. The intuitive thought (*matis*) emerging from 'within', i.e. from within the heart, follows, or 'joins', the 'gift of the cow', which is the ghee or milk oblation. The poet's thought and his offering blend together in the ritual.

The sacrifice is equivalent to the Cosmic Order: *yajña* = *ṛta*; the establishment of the sacrificial rite

The word *ṛta* is often used throughout the Ṛgveda in the singular with the significance of rite, or even with both meanings of law and rite. J. Gonda remarks 'Another "seat" or "manifestation" of ṛta is the recurrent ritual for without contending that "ṛta" and "sacrifice" are completely synonymous . . . the regular rites were intelligibly enough regarded as representing the fundamental principle of ṛta.'[44]

In the following examples the sense of ritual is possible, together with that of harmony or order:

> Let not the lustful enter our *ṛta* [the enclosure of our sacrifice].

> By *ṛta* I purify both heaven and earth [by the sacrificial rite I purify Heaven and Earth].

> May this *ṛta* protect me for a hundred springs [May this sacrifice which I am now performing protect me].

> I sing to Heaven and Earth and offer (this) *ṛta* [I offer this sacrifice, etc.].

Agni, 'eye and guardian of mighty *ṛta*' (order) is identified with Varuṇa 'when thou approachest the *ṛta*' (ritual). Some verses telescope into the succinct expression of *ṛta* a vast process of thought:

> Full of understanding, I have first uttered this *ṛta* to Heaven and Earth (for all) to hear.[45]

The sacrificial rite, in the poets' view, performed by man in commemoration of the primordial divine sacrifice, becomes equivalent to the Cosmic Order which is the manifestation of the divine sacrifice. Both *ṛta* and *yajña* signify a process going on according to an inherent law of harmony.

The ever 'renewed' song of the ancient vision abounds in the norms of the Cosmic Order. It therefore proclaims the truth. If the song (which incidentally may be a symbol of the whole ritualistic order) did not conform to the inherent structure of things, if it was not an epitome of it, it would fail in its purpose. So the poet, understanding these things full well, can claim 'I offer this *ṛta*', the *ṛta* including and epitomising everything. So

> the visionary-insight has shone from the seat of *ṛta*

and

> the sun illuminates by means of the song born of *ṛta*.[46]

The ancient fathers thus 'fostered the *ṛta*', the sacrificial rite as a mirror of the Cosmic Order, as they

> gained their bright inspiration whilst singing their praises.[47]

The primordial sacrifice was established by the gods. This we know from the famous *Puruṣa sūkta* (X.90). As primeval sacrificers

> they have fixed the year, month, day and night, the ritual and the holy word (VII.66.11ab).

Thus was the sacrifice given the seal of godly statute:

> the gods, denizens of heaven, flame-tongued, promote the *ṛta* as they sit brooding in the source of the *ṛta* (X.65.7ab).

The sacrificial ritual became established on earth by the patriarchs as a religious institution expressive of the Cosmic Order and man's active participation in it. Its performance in all its details was worked out so as to be in harmony with the seasons and the planetary movements.[48]

> They established the *ṛta;* they maintained its realisation (I.71.3a)

literally 'they set in motion the vision of it'. The establishment of the *ṛta* as the norms of Cosmic Order includes social, ethical and ritual order. Thereby

> They made for us a way to lofty heaven (I.71.2c).

This way is the institution of the 'rites of sacrificial worship' whereby all come together to pay homage to the gods (X.67.2) and commune with them.

> stating the ṛta visioning aright . . .they first thought out the forms of the sacrificial rite (X.67.2cd).

This is the human response to the revelation of the Cosmic Order and the human attempt at re-enacting it in its own microscopic and symbolic version. In 'instituting the *ṛta*', the patriarchs 'stimulated its insight',[49] they

> marked out the ancient unmeasured spheres.[50]

But they act in accordance with a higher order which finds itself reflected at all levels of Vedic cosmogony:

> I worship the *ṛta* regulated by the *ṛta* (IV.3.9).

This verse admits the existence of a higher order in which is rooted all other orders – the divine as well as the human in the sacrificial rite.

> By *ṛta* is *ṛta* firmly fixed there where they unyoke Sūrya's horses (V.62.1ab).

By *ṛta,* the law of harmony, as it is in heaven at the transcendental level, is *ṛta,* its reflection both in the ethical and ritualistic domain of men's endeavours, firmly established.

A further implication is taken up by H. Aguilar who argues that, firstly,

> The sacrifice is not merely a cultual part of ṛta . . . but the very same ṛta, conceived as the primordial source from which originates the cosmos. . . . In order to become not just an intermediate principle but the first principle or cause of the world it is a condition *sine qua non* for the sacrifice not to end in the gods but to be above them in all respects.

Secondly, that in religions the sacrifice consists always of a gift to a higher Being, but in the Vedas 'the sacrifice is not ordained to something different or higher than itself, but at the most to a higher form of itself'.[51] The idea of the celestial order as the blueprint of all orders is confirmed in the following:

> By means of *ṛta,* with the power of sacrifice, they (the patriarchs) upheld the all-supporting *ṛta* in the highest heaven (V.15.2ab).

This is the supreme *ṛta* of which the natural order and the human rite are reflections. As explained by Aguilar, 'the pitṛs do not fail to maintain in heaven the same sacrificial order they had maintained on earth and they do it by virtue of a sacrificial order which is undoubtedly superior.'[52] This is the Cosmic Order which we may consider as the archetypal Order of which every order is a fragmented reflection. By means of the ordered, ritualistic expression of the universal oblation established on earth as a token of the heavenly *ṛta*, the patriarchs helped men to reach out to a loftier vision and code of life. By such paths of vision they found

> What had been kept secretly hidden as the supreme domain of the sacrifice.[53]

The whole process of the institution of the sacrificial rite, its intimate connection with the visionary insight or revelation of the Cosmic Order, its 'thinking out' and establishment, its ultimate prize, immortality,[54] is outlined in these allusions to the patriarchs' feats. The *yajña* was considered the most effective means of conveying to man the meaning of the Cosmic Order and of representing it in a concrete and dramatic form in which man could himself play a vital role. Hence *ṛta*, in one of its connotations, came to mean the ordered expression in rite of the universal oblation, the give and take, the receiving and the returning, the eternal rhythm of the orderly movement of the universe. So the wish expressed in an invocation to Mitra, Varuṇa, Aryaman

> May we be champions of your *ṛta*[55]

evidences the ṛṣi's aspiration towards the full accomplishment of the *ṛta* and his total participation in it.

As already shown in IV.3.9 where the poet solicits

> the *ṛta* regulated by the *ṛta*

the ṛṣi invokes the Cosmic Order as it is reflected in the ritual, man's symbol of it, and through the latter transforms his invocation into a sacred action, a symbolic return to the universe of what the universe has given him. This verse consists of two parts which to all appearances seem unrelated. It could be rendered loosely thus:

> I solicit the truth established by the sacrificial rite; even so O Agni, just as the raw cow gives of her sweet, ripe milk (do thou perform thy task of transmuter) (IV.3.9ab).

The image in the second part – pointing to the natural process whereby the cow, though raw, is yet able to give sweet, ripe milk, a 'mature' offering, the means of the transformation effected, incomprehensible to man, being inherent in nature and an expression of the law of harmonious becoming, *ṛta* – surely has a bearing upon the first part of the verse: just as the milk is the end product of the harmonious working together of natural processes – the cow gives that which has been transformed through her – so *ṛta* as the rite is the mature human expression and symbol of the harmonious working of laws as summed up in the Cosmic Order, and man's response to it in his moral life and the ethics of society. The cow could here personify the *ṛta* in both its meanings.

A similar thought, touching upon the interlinkedness of all things as they proceed in accordance with the inherent law of their being, is expressed as 'butter and milk . . . drawn from the essence of the cloud', whence *amṛta*, the nectar of immortality, is 'produced as the centre of *ṛta*'. (IX.74.4ab). Through the bond of the sacrifice performed in accordance with the norms of Order, Soma (he who grants *amṛta*) is pleased and causes rain to fall from the sky (IX.86.14c); nature thereby fructifies; so butter and milk, the essence of nature's gifts, are extracted from heaven. The 'soul-infused cloud' (*ātmanvan nabhas*), however, may be considered the effect of man's offering. For just as the bee collects the pollen of the flower to transform it into honey, just as the cow gives its essence – milk – to feed its young, so man offers in the sacrifice that which he has assimilated through the process of purification. The very core of *ṛta* wherefrom emerges the nectar is this sacrificial ritual, symbol of universal and continuous exchange: the transformation process wherein man gives up those desires which constitute his earthly self for those which link him to the gods or his godly self, thereby strengthening his bond with heaven and contributing to the higher aspect of the law of give and take. Hence the Vedic poet summed up this mighty interconnection in one of those bold and far-reaching statements which to the superficial understanding may appear unintelligible:

> To law, wide Earth and deep (Heaven) – supreme milch kine – yield back their milk (IV.23.10d).

The purifying function of the sacrifice

In addition to serving as a spiritual experience and a social participation, the ritual had an expiatory function:[56] the re-enactment of the cosmic process was considered a purifying action. The verb √pū, 'purify', is used with both the word *yajña* and *ṛta* as reflected in the sacrifice. Thus:

> By *ṛta* I purify heaven and earth (I.133.1a)

i.e. by means of my participation into that orderly process which I re-create in the ritual as a mirror of the cosmic order, I bring about a purifying effect in the two worlds of spirit and matter. It is a conscious return to an original harmony, a going back to the root, the ground of all being, truth, *satya*.

Performing the ritual the poet proclaims:

> This by the holy syllable do I imitate.
> I purify in the centre of *ṛta*.[57]

This links the sacred function to the essence of *ṛta*. Reality, in its aspect of ordered activity, *ṛta*, is an expression of harmony. This is the truth (*satya*) on which all things are based;[58] in this central point of truth is found that which has neither beginning nor end, that which *is*: from that centre originates the nectar of immortality (*amṛta*).[59]

The purpose of the sacrifice (*yajña*) viewed as a homage paid by man to deity not only purifies but establishes a specific relationship between earthly and divine powers. The gods show men the way. To the Maruts

> To you, the pure, I send forth a pure sacrificial-offering.
> By *ṛta*, they observing the *ṛw10ta*, came to truth (VII.56.12).

Prayers addressed to Agni, the high priest of the ritual, show him as the purifying agent *par excellence*:

> Burn up our sin away (I.97.1a)

> O Agni, with thy flame outspread as a filter, purify this our invocation.[60]

Even the evocative word has to pass through the sieve of the flame so that no dross may remain ere it reaches the gods. Can such a purified word be solely a demand for worldly goods? Borne by Agni it must 'shine in heaven'.[61] The kine asked for have frequently a double meaning; they refer to those herds of the heavens, those rays of light that illuminate both the universe and human consciousness.[62]

Soma, the 'father', like Agni, is asked to

> purify us with (his) filter (IX.67.22a).

Three filters are said to be set within his heart.[63] Each of these may apply to one of the three levels of human experience: physical, emotional, psycho-mental.[64] Whatever is received from the innermost depth is filtered before it reaches any of the levels. But since Soma is begged to 'purify us with (his) filter' the purpose of the filter is self-evident. That the various references to it go far beyond the mere physical instrument used in the ritual is even borne out by the mention that Soma is distilled with the 'efficacy of true speech' (*ṛtavākena satyena*), 'with tapas, with faith' (*śraddhayā*).[65] In other words, the true offering from which results the true Soma, or enlightenment, can be effected only through pure contemplation, spiritual exertion, that state in which the divine creative fire is aroused – *tapas*. But this, we are reminded, can only be achieved through faith and the true invocation which is the keynote of the life, i.e. in all sincerity. For the invocation can only be true if it springs from deep within the heart, a heart purified of all the lower inclinations that flesh is heir to. This verse reveals the Vedic understanding of sacrifice as not only an orderly rite, a mechanical mirroring in miniature of the eternal circuit which makes the cosmic process, but a heartfelt offering of what is best in man, dependent upon the values of the 'heart'[66] and resulting in deeper understanding which itself is bound up with the inspiration that plays a most important part in the sacrifice.

In an invocation addressed to Vāyu the wind, Soma is

stated to be pressed by men who have cleansed themselves from sin (I.134.6). This could be a hint that the ritual preparation of the Soma would require a prior purification. How that prior cleansing has been effected is not told. But in another hymn the generous gods (Vasus) are asked:

> Make me clean by means of the vision (*dhiyā*, IX.67.27),

or vision-converted-into-song, perhaps 'revelation'. The function of the latter[67] is thus not only a means for the poet of entering into contact with other realms, fathoming their essence and expressing it by means of words, but this penetration, this contact with unseen powers through supranormal insight, is thought to have a wholesome effect upon him, to be purifying; or perhaps he cannot enter into that state of vision-inspiration unless he has previously undergone some kind of purification. The vision or inspiration (*dhīḥ*) is also shown in a prayer addressed to Heaven and Earth as capable of removing sin:

> Whatever transgression we have ever committed against
> the gods, against a friend, or against the family head,
> may this our dhīḥ act as (be) expiation (I.185.8abc).

Visions-translated-into-songs or revelations thus do have a ritual and purifying function.

According to the traditional order which has become the path of *ṛta*, the inspired revelations are offered up to the gods:

> Along the path of *ṛta*, from the sacred-work, speed forth
> the inspirations to Indra-Agni.[68]

By their means the ṛṣi will be able to enter into communication with the devas. This is the sacred work (*apas*). The sacrifice helps the intuiting of the vision and the latter helps the right process of the ritual and its furtherance. It enables man to grasp in its essence the *ṛta* or reality in its ordered activity.

THIRTEEN

The role of Soma and Agni in the ritual

Introduction

Two gods are intimately connected with the ritual and the purification process of the sacrifice: Soma and Agni. Each one plays as dynamic a role as the other. Both purify, inspire and exalt. Both are different and yet two facets of the one power that uplifts and opens up the vision divine. The one purifies by means of the filter set in his heart, and grants ecstasy-illumination-immortality, the other's filter is his all devouring flame which 'raises the mortal to highest immortality'. Both stimulate mental activity and spur on to exaltation. Indeed their functions overlap considerably. Soma's function seems to be that of an introvert, Agni's that of an extrovert within the *ṛta*. But they meet in their effect: purifying, exalting, illuminating, the two become one in the transmutation process which enables man to enter into a supersensuous realm which is perhaps one of the aims of the sacrificial ritual. Both can be thought of in terms of the consumer and the consumed, the offerer and the offered, the devourer and the devoured. Agni, the high priest (*purohita*) of the rite offers up the oblation transmuted through his ministry; the response is the exalted vision of godhood granted by Soma, the king. Both take the worshipper to that innermost centre whence the state of 'immortality' is experienced. As beautifully stated by S. Kramrisch:

Agni, Fire, the spark and flame of life, and Soma, the

> elixir of life, descend from heaven to earth and ascend from earth to heaven. Their track of liquid fire extends from here to there and from the beyond down to the earth with the systole and diastole as it were of inspiration and aspiration. The place where each of these gods comes to earth and whence he ascends is the place of the sacrifice, the sacrificial altar, the navel of the world. . . . The rites of the sacrifice confirm and celebrate the tracks of these two gods which traverse the cosmos and the inner world of man.[1]

An analysis of the function and meaning of these two deities in the ritual may serve to clarify certain obscurities in the significance of the Vedic sacrifice and its part in the vast cosmic order as envisaged by the ancient Vedic sages.

The myths concerning Soma and Agni

The celebrated myth of Soma's theft from the heavens, the earliest narrative of which is found in Ṛgv. IV.26 and 27, relates how the swift-flying eagle on which Indra seems to be mounted (IV.27.3 and 4) is said to have flown to heaven where Soma is hidden and closely guarded, in order to bring it down to earth,

> to bear to man the oblation loved of the gods.

From far aloft did the straight-flying bird bring the exhilarating plant, screaming down from heaven as the archer, guardian of Soma, aimed his arrow and there fell a feather from the eagle, who nevertheless returned triumphant with the plant of immortality in his talons.

Some[2] have identified the eagle with Agni the lightning and the Soma with rain! Various altered versions of the Ṛgvedic myth are found in the later literature; in the Brāhmaṇas it is the *gāyatrī,* the famous metre which belongs to and is identified with Agni, that flies to heaven, after the other metres, the *jagatī* and the *triṣṭubh,* have failed in the same mission, and succeeds in wrenching the Soma.

Parallel myths may be found in the Avestan, the Scandinavian mythology where Odin as an eagle steals the mead of immortality from Suttung, and the Babylonian mythology where Etana ascends to heaven on an eagle in search of the plant of birth. In each case, he guardians of the treasure discharge their weapons but do not kill the transgressor, though the latter suffers the loss of something of himself – a feather as in the case of the Ṛgvedic eagle – and thus pays the price of intruding upon celestial mysteries.

Soma is reputed to grow on the mountain; 'the mountains nourish him'. But mountains are sacred abodes that sooner or later, in the mythopoeic mind, become fused with heaven. So he dwells 'on the third ridge, in the bright realm of heaven' (IX.86.27). As a plant he is found on the mountain, as a god he abides in the highest heaven (*parame vyoman* III.32.10). The purpose of snatching the Soma from heaven is thus explained:

> For all to see the universal light, the bird brought us the guardian of *ṛta*, the cleaver of the heaven (IX.48.4).

The two parallel happenings, the descent of fire and the descent of Soma, are brought together in one verse:

> One of you Mātariśvan brought from heaven;
> the other the falcon tore from the mountain (I.93.6ab).

As rightly stated by Wendy O'Flaherty

> Agni and Soma are linked in many ways. As fire and liquid, they are complementary oppositions that unite in the Indo-European concept of the fiery liquid, the elixir of immortality. . . .
>
> Agni and Soma are the two sources of the inspiration that enables the Vedic poet to find and understand the meaning of the sacrifice and of his life. They are different sorts of inspiration: Agni is Apollonian, explaining the sacrifice; he represents the cultivated, cooked, cultured aspects of ritual. Soma is Dionysian, explaining the vision of life; he represents the wild, raw, disruptive aspects of ritual. The Vedic sacrifice embraces them both.[3]

The myth that concerns Agni (X.51) tells how he tired of his task as priest of the sacrifice and so just disappeared; fled away hiding in the waters. The gods, one-minded, track him as one does a 'thief lurking in dark cave with stolen cow' (I.65.10), seeking him out in the plants and the waters; therein Yama espied him concealing himself, Yama the original man or primordial Man who chose to be mortal. So, a 'mortal-immortal' discovered Agni in the highest seat, the source of the cosmic waters, the lap of *ṛta*. Yama found Agni by virtue of their identical nature, for Yama as god of death dissolves all forms just as Agni burns all dross away.

Indra, the knower, the first to possess mind (*prathamo manasvān*, II.12.1), the meditating god (X.32.1), also found Agni, as stated elsewhere (X.32.6), lying in the waters. With Agni is the power, creative and illuminating, which takes 'the mortal to highest immortality' (I.31.7). Finally, the holy ones, the gods

> found the thrice seven secret words hidden in thee,
> whereby, conjointly, they protect the immortal one
> (I.72.6).

These thrice seven secret words, though hidden in Agni, are the names of the Mother (IV.1.16), of Nature in all her infinite variety. 'They are', says Stella Kramrisch, 'the total symphony on the three levels of Fire of the sevenfold melody in the beyond. With Agni found, the three times seven secret words were found.'[4] As she points out:

> The secret flight of Agni, the Fire . . . does not take place in time, but in principle, in the archetypal age, where the pattern was established and is valid at all times.
>
> The Light and Life that is in the Godhead goes over to Indra, the god . . . who creates this world. All this happened in no-time, in the beyond; and while the Fire that is latent in the firewood is kindled, whenever it is being kindled this Fire which burns here on earth goes back each time and comes from the source, Ṛta, and the godhead.[5]

Such a myth is concerned with the search by the mortal for the immortal flame hidden deep within the mortal, in the ocean of the heart (the space of the heart of the Upaniṣads), in the 'house' or human tabernacle, that flame that holds all power, the transcendent light without which nothing that is done can be done, the 'god-given treasure' (IV.1.10 etc.) which the patriarchs discovered on 'awakening', the secret of which gives man insight into and power over the manifested world.

Soma: the ancient kavi, the inspired and inspirer

Two points of importance have to be examined with respect to Soma: Soma as a 'knower', the inspired-and-inspirer (*kaviḥ*),[6] hence the visionary; and Soma as the essence of the *ṛta*.

Soma in the Ṛgveda is a plant (probably *asclepias acida*) that, when pounded and mixed with milk, yielded a juice which the poets describe as the 'golden drops', which had an intoxicating effect and granted ecstasy. Soma is the 'fiery juice', the flame drink about which the poet exclaims: 'wound not our heart with dazzling flame' (VIII.79.8). The main purport of the hymns addressed to Soma is not only to extol him but also to sing of this exaltation. In the Ṛgveda Soma is not identified with the moon, except in a few late hymns.

There is a constant fusion and perhaps deliberate confusion, in the figure of the god Soma, of the objective and the subjective – the plant and its juice as the object of the poet's care, and the ecstasy produced in the poet as he partakes of the drink. A very clear hint as to a hidden meaning to Soma is found in the great bridal hymn:

> Soma is thought to have been drunk when they press the plant. The Soma whom the Brahmans know, of that no one tastes.
> No earthly born can taste of thee (X.85.3,4).

There are here two meanings to Soma; the plant that yields

a state of ecstasy, and the god Soma whose meaning goes beyond that of ecstasy and whom mortals cannot know. But through that ecstasy the mortal enters into a dimensionless state which, being beyond time and space and any conditioning, to him is 'immortality'. Only few can realise this, and therefore only few can know the real meaning of Soma. Being 'twice-born' or born anew, alone the true Brahman can know the real significance of Soma.

Soma is said to be born in the 'womb of *ṛta*', the sacrificial ritual as it mirrors the cosmic process; Soma thus embodies the norms of cosmic order, and being the exhilarating drink that bestows enlightenment and immortality he inspires the sacrificer, uplifts him to the domain of the gods and grants him that insight which pierces to the very core of the universe, hence of the cosmic order. He is the illuminator *par excellence*, the one whose action expresses the *ṛta* in its deepest sense; the one who is asked to 'further the lofty rite';[7] that is, through his guidance which is illuminative insight, the ritual will be a mirror of the cosmic process. In this request his role in the ritual is declared essential. As he 'flows forth purified', he, like Agni, brings to the altar those 'fosterers of the cosmic order' who are the gods. This may also refer to the powers of inner perception opened up by Soma which allowed the poets to come into contact with supra-sensory beings, to become aware of another dimension to life. This may be the meaning of the following:

> he, thundering, has produced the gods (IX.42.4)

i.e. has made the divine powers manifest.[8]

Called the 'embryo of the *ṛta*', he 'engenders the *ṛta*' and 'grows mighty by *ṛta*', and is identified with the 'mighty *ṛta*' in both its ritual and cosmic sense:

> He born in *ṛta*, waxing strong by *ṛta*, king, god and mighty *ṛta*.[9]

> Swiftly he flows, as lofty Cosmic Order, to the purifying sieve.[10]

His juice flows into the sieve according to those immemorial practices of the sacrificial ritual as they had been handed down by the wise, hence in accordance with the traditional wisdom.

Such expressions as 'womb of *ṛta*', 'seat of *ṛta*', whenever used in connection with Soma (or Agni), are evidently referring to the enclosure where the seers congregate to perform the rite and await the manifestations of the visions which, with the help of Soma, enable them to pass beyond the normal range of perception, to expand their insight quite beyond the power of the intellect:

> In the seat of *ṛta* the seers keep watch over their intuitive-insight, luminous-as-heaven and bright-like-lightning,
>
> Men urge Soma forth to the sacrifice with all-expressive vision.[11]

The maintenance of the law as reflected in the sacrificial rite and the illuminating action of Soma are both functions of the god which are activated by means of the ritual. That Soma the 'inspirer', or divine-poet (*kaviḥ*), is said to be generated by means of purposeful intellectual activity (*dakṣena manasā*) presupposes a thinker, whether human or godly, to make Soma manifest and goes beyond the mere physical action of crushing the plant for its juice:

> Through will and mental activity is generated the inspirer placed as embryo of the *ṛta* beyond the twins (IX.68.5).

Is there a hint here as to the inherent meaning of Soma, namely that illumination which flashes forth or is 'generated' in the worshipper when he has duly performed all his duties and concentrated his mind on higher purposes, visioned aright and invoked the presence of the gods in the right way? Elsewhere (I.46.8) the Soma drops are said to be prepared by means of the visionary insight (*dhiyā*) of the poet. Are these 'drops' a technical term referring to the culminating point of an experience preluded by the visionary insight, in fact a deepening of that revelation which takes the seer into a state of ecstasy such as Soma produces?

The mention that Soma is placed 'beyond the twins' is also

worthy of observation. The word 'twins' may refer to Heaven and Earth, or to any duality of the manifested worlds, or of man. His origin is beyond the worlds of complements or of opposites. In the case of a man illumined by Soma, this expression would refer to the man's ability to enter a world beyond the opposites, hence the mystical state.

Soma's action upon the seer is dual; he opens his eyes and loosens his tongue, i.e. grants him that eloquence that makes him a poet, 'golden-tongued' as well as 'sun-eyed':

> This here when drunk arouses the word, this has awakened the eager intuitive insight (VI.47.3ab).

As 'arouser' he 'unseals the intuitive-perception of the priest'.[12] Visioning and singing are the essential qualifications of the ṛṣi; these are enhanced by Soma, the ecstasy-granter, who

> releases the mind-yoked vision even as the thunder releases the rain (IX.100.3ab).

The mental activity which generates him is thus linked to visioning. He is son of visionary insights who himself grants insight or illumination. Invoked as the inspirer, he reveals the part he plays in, and the purpose of, the sacrificial rite: in enlightening man he enables man to penetrate into the very fabric of the cosmic order, he grants man the vision of *ṛta*. The eyes that can visualise this order, the mind that can comprehend its full meaning, can only be those enlightened by Soma. They are his gifts to the ṛṣi. Hence he is the 'treasure-granter' (*ratnadhā*), who seated 'in the womb of *ṛta*' is the very 'soul of the sacrifice',[13] the very core of the worship.

For such reasons Soma is proclaimed the 'lord' of the poet's 'vision',[14] the 'lord of unalloyed speech',[15] the 'truly inspired one', or real poet (*ṛtaḥ kaviḥ*) (IX.62.3). Since he makes the vision of *ṛta* perceptible he is most ṛṣi-like (*ṛṣiṣāḍ*):

> He flows, king of all celestial seership, roaring forth the vision of *ṛta*, surpassing the ṛṣi (IX.76.4ab).

This supreme 'vision of cosmic order' coming to the ṛṣi through Soma in the ritual is a transcendental experience that surpasses the ordinary mental faculty. The whole ritual is implied as helping to bring to consciousness that which is deeply buried in the ṛṣi. This is the

> glory for which Soma is born

that he should

> give glory and vigour to the singers (who) donning (his) glory have obtained immortality (IX.94.4).

The seer when exalted enters into a state of ecstasy which is Soma's gift, hence he dons Soma's glory, sees and knows through Soma's eyes.[16]

Soma is furthermore addressed as Brahmaṇaspati, the Lord of prayer (*brahman*), of that power of the utterance and invocation which blends appeal and command:

> Outspread is thy purifying filter, O Brahmaṇaspati (IX.83.1a).

The purification opens the way for prayer, or direct communication:

> Whilst being purified Soma finds a pathway for prayer.[17]

The very purifying and enlightening role of Soma, the power of ecstasy which he grants, blazes the track for that prayer which will, by its invocation, bring gods and men together in communion. In this essential aim of the sacrifice we glimpse again the reflection of the cosmic order, the *ṛta* which implies the bringing, the working together in orderly sequence of the various forces of the world, for holy creative purposes: the 'renewal' of those who participate in the drama, hence the renewal of the whole creation.

Through the impact of inspiration and visioning the seer bursts into eloquence. And this serves to exalt the *ṛta*: the inherent harmony in accordance with which all things move is visioned as *satya*, truth. To proclaim this truth is Soma's glory, his act of faith; such words as order, truth, faith brought together in a verse that describes Soma, the exalter of the sacrifice, reveal the essential meaning of the ritual:

> With the right word (*ṛtavākena*), with truth (*satyena*) and faith and fervour (*tapas*) is (Soma) pressed (IX.113.2).

The true Soma which grants illumination cannot be known unless the seer reaches a state of purity pregnant with faith, with fervour of heart, with sincerity. Only then will the power of the word be discovered and the illuminating vision and its eloquent expression be forthcoming. These verses lift the veil that hides that side of the sacrifice left impenetrable to Western scholars; the pressing-and-pouring of the Soma is not only a physical act, but an act of faith and meditation expressive of true prayer, reaching out far beyond the mere physical action, and Soma merges into man as the mouthpiece of man's tribute to the divine powers. This inner meaning was completely ignored, and as a result the Vedic sacrifice was dismissed as meaningless 'hocus pocus'.[18]

So the priests

> have assigned him (Soma) as messenger for the sake of probing the divine-insights (IX.99.5cd).

Through Soma's illuminating function the seers are able to grasp the meaning inherent in their intuitive insights; they are able to penetrate into the significance of the Cosmic Order. So they ask Soma to 'increase' the power of the 'word' and 'engender abundance' of ideas, or visions (IX.97.36).

Soma's three domains

Soma's three domains are referred to in several verses but remain on the whole a mystery to Western exegesis.

> Three are the utterances to which he gives voice, he, conveyor of the cosmic vision, the intuitive perception (manifesting) brahma.[19]

The *'conveyor of the vision of ṛta'* declares the function of Soma; he opens the inner eye to the perception of the cosmic order. The fact that two of the keywords, 'intuitive perception' and the 'invocative power of the word' (*brahman*), are juxtaposed to the 'vision of *ṛta*' points to the sacredness of the whole procedure and to one of the several meanings of the sacrifice, namely a means of coming into communion with extra sensory powers and addressing to them man's higher aspiration. J. Gonda's interpretation of this verse may help in clarifying its meaning:

> Soma is said to initiate the process of coming, by way of a visionary intuition in touch with the norms, 'laws' and structure which underlie the phenomenal world and make it an organic whole and to set also an analogous mental process with regard to brahman in motion.[20]

The whole purpose of the sacrificial ritual is here set forth; hence the identification of *ṛta* with *yajña*.

The three voices or utterances which J. Gonda ascribes to the recitations of the priests, the lowing of the cows and the sound of the soma[21] may have a far deeper meaning, for they may pertain to the three domains or habitations (*dhāma*) of Soma: the realm of sound, of light and of supreme illumination, each being an expression of Soma's action. Eloquence or the power of the word, i.e. the realm of sound; enlightenment or light, i.e. the vision of truth; and full illumination or immortality, i.e. pure *vidya* or wisdom – these are his gifts to the ṛṣis, his sphere or domain of activity. His threefold estate, and even a fourth one, are mentioned:

> Far-seeing mind, enlightening, light bestower, skilled in a thousand ways,
> leader of the inspired-poets, the mighty one strives after his third estate.[22]
>
> The mighty one declares his fourth estate.[23]

Are these three (or four) estates, or domains or spheres of action, or attainments of Soma, expressive of the states of consciousness which were to be analysed in the Upaniṣads, but must have been known much earlier from the evidence of the Ṛgveda? Degrees of enlightenment in the inner path to supreme illumination of which Soma was personified as the inspirer?

The first domain: sound, *vāc*

That sound, or perhaps what we might now call the primeval resonance, may be an all important domain of the ṛṣis' probing insight, a domain vivified by the action of Soma, is brought to our notice in the ṛṣis' search for the divine word, *vāc*, during their act of worship (*yajña*):

> They tracked down the path of *vāc* by means of the sacrificial rite (X.71.3ab).

The ṛṣis personified *vāc*, the word, or original sound which the famous hymn to Vāc (X.125) tells us is the fountain source of all manifested life, as the Queen (*rāṣṭrī*) holding all worlds together (X.125.8). Following the path to Vāc, the primordial sound-vibration-motion, they discovered her 'abiding' among the ṛṣis, i.e. themselves. Vāc is the power of the word hidden in the seer and aroused through the rite, by means of Soma. As R. Panikkar says:

> The Word is not only speech, though constitutively connected with it; it is also intelligibility, the principle of reason, the power of the intellect, the rational structure of reality.
>
> The Word is not only sound, not only idea and intelligibility; it is also action, spirit, the unique Word permeating everything.[24]

In the *word* is not only the sound, but also the principle of ratio, rational construction and rhythm, balance.[25] This creative power is said to dwell in the highest heaven (I.164.34). It was expressed in the Greek word *logos* and is also given pre-eminence in the Christian scriptures:

> In the beginning was the Word, and the Word was with God, and the Word was God (John 1:1).

The Vedic sages discovered that every state of consciousness has its own inherent vibration and every vibration necessarily affects states of consciousness. Particular states of consciousness can thus be brought about by initiating particular kinds of sounds. We know, through music, how certain kinds of music or rhythms are soothing, uplifting or exhilarating, others irritating or depressing. Through their investigation in the realm of sound the ṛṣis reached back to the *anahata śabda*, the 'unstruck sound', the sound of creation.

> Having given shape to her they produce her variously (X.71.3c)

i.e. they weave their hymns around this basic sound potency. This would indicate a knowledge of the mysteries of sound and its impact and effects.

The sound our ears register is only the coarsest manifestation of sound. As the Ṛgveda hints, the divine Vāc has been

> meted out in four sections which only the wise Brahmans know (I.164.45).

Three of these sections are hidden from human ken; that which man hears is only the fourth or lowest, each one corresponding to a particular level of manifestation.

> Three parts, hidden in deep secret, humans do not stir into action; the fourth part of Speech is what men speak.[26]

For this reason the ṛṣi Dīrghatamas declares that only when visited by the 'first-born of holy-order' does man gain insight into *vāc*, and obtain a share of her, an insight into her mystery and thereby the power of the word. Thus only through the divine influx, whether caused by Agni or Soma as the 'first born of *ṛta*', does man become conscious of the potency of the creative word and able to use it to fruitful purposes.[27]

For the ṛṣis the whole process of worshipping meant to set up an inner light by means of chanting. They conceived light and sound, shining and singing as interchangeable. For them to sound the right word was to emit a light, hence sound manifests as light and vice versa, and sound-light attracts the denizens of the worlds of luminosity – the devas. The *arka* is a ray, a flash of lightning as well as a hymn and praise. Songs 'glow with the force of fire' (X.68.6). Agni is invoked for *brahma*, the evocative power of the word ('Bring fruitful prayer O Agni', VI.16.36), that the latter 'may shine in heaven'. The holy word shines. This is the ṛṣis' prayer, the blending of sound and light in a powerful invocation.

And so the invocation goes:

> Proclaim three light-projecting-words (that) milk the honey-yielding udder (VII.101.1).

The illuminating power of the word draws down, or milks, the nectar of heaven. The divine vision-speech is the celestial cow which the bards milk in a hundred streams.

> Some singing, conceived the great chant whereby they caused the sun to shine (VIII.29.10).

Since Soma gives eloquence to the poet, he is the 'tongue of the *ṛta*' as that eloquence expresses itself through man.[28] As J. Gonda explains, 'Soma . . . the speaker and lord of this "vision" . . . imparts the vision to the ṛṣi's mind . . . the ultimate source of the visions is the *ṛta*, the god Soma transmitting them to the ṛṣis by the instrumentality of his tongue.'[29]

The second domain: vision, illumination, revelation

The second domain, so intimately connected with the first, the realm of light to which Soma led the ṛṣi, was described as sunlight, or celestial light:

> Visions vie for the sunlight as people (IX.94.1b).

Soma's potent action upon the mind, shown as 'stimulating' in the bard

> the vision of *ṛta* for the sake of the sacrifice[30]

wins him the title of 'lord of mind',[31] 'all-knower', the 'far-seeing, wise one';[32] that is, he fosters that insight that permits the bard to penetrate into and understand the laws of the universe:

> knowing the *ṛta*, he, foremost, wisest, celestial lord, was indeed chosen to proclaim it (IX.70.6).

Through his intelligent perception (*kratvā*) and 'bright eyes' he opens out 'the stall of heaven', thus 'releasing the vision of *ṛta*':

> Through thine understanding, with thy bright eyes, thou hast opened out heaven's stall, fostering the vision of *ṛta* for the ritual (IX.102.8).

The reference to 'bright eyes' is characteristic of Soma whose drops arouse, or bestow, an intuitive perception which as visionary insight is said to 'shine from the seat of *ṛta*' (X.111.2a). Soma's drops foster wise perception, they find celestial light for the seer,[33] for Soma

> Seated in the womb of *ṛta*, perceives by means of heaven's light (IX.13.9b).

Intoxicated with Soma the seer is enabled to penetrate or see 'heaven', which otherwise would be shut out of his vision. 'Far-sighted, brightly-he-shines', 'thousand-eyed', made 'sun-eyed' like the gods through the action of Soma, the seer is able to delve into the mysteries of the beyond.

In that state of heightened awareness the Vedic bards admitted to beholding the 'golden one' not with visionary thoughts or mental insight but with the 'very eyes of Soma (our) very eyes'.[34]

The close link between Soma and the all-seeing sun, the 'overlord of sights' as the Atharvaveda (V.24.9) qualifies Sūrya who, in the Ṛgveda, is called the 'wide-visioned one',[35] is a pointer to Soma's illuminating action in man. He is said to 'generate the sun', to 'place the eye with the sun', that is, to illumine human consciousness, to 'produce' Sūrya

> midst mortals for maintenance of the law and for prized immortality.[36]

This, together with such statements as

> Having received from my father the vision of *ṛta*,
> I was born like a sun (VIII.6.10)

is one of the most important verses relating to the sun and to the Vedic vision in so far as it shows Sūrya as symbolic of that illuminating principle elsewhere aroused through Soma's action. The end of illumination is the understanding and the maintenance of the universal law in all its aspects and the entering into that state of ecstasy beyond the opposites, or immortality.

Mythological expressions such as Soma mounting on Sūrya's chariot, Soma mounting up with Sūrya to the sky, Soma making Sūrya ascend the heavens, Soma shining by means of the sun, Soma 'donning the rays of Sūrya',[37] all tend to indicate one overwhelming state of exaltation culminating in illumination that is akin to the solar splendour to which Soma gives birth. The bond between Soma and Sūrya in the sacrificial ritual hints at a close connection between the sacrifice and the illumination. Having

> wrapped himself up with Sūrya's rays, he (Soma) spins the triply twisted thread, as he knows, conducting the new operation of *ṛta* (IX.86.32abc).

In a pithy summing up of the ritual, this verse indicates Soma's action which, in conjunction with that illumination described as solar, spins the thread of the sacrifice, that is, relates the two worlds of the manifest and the unmanifest, thereby 'conducting' the norms of the cosmic order. The action of the ritual is often compared to a thread spinning out a web,[38] the image reminding us that the sacrifice was viewed as a mode of action whereby man himself, so to speak 'wove' the pattern or web of the cosmic action, the warp of *ṛta* being spun in the 'sieve'. Thus the action of men and gods is closely interlinked; man sets the warp through the grace of the deity and in his spinning makes his contribution to the cosmic order. This is a very ancient action that dates back to the origin of the world and is 'well known'. Indra is thus invited to

> spin out the thread of ancient time, as is well known.[39]

Soma, essence of *ṛta*

As already seen in several instances,[40] Soma holds in himself the extremes of manifestation and in this respect is a worthy embodiment of *ṛta* which holds all opposite tensions in equilibrium:

> Shining over both worlds, like a darting deer, thou rushest forward seated in the womb of *ṛta* (IX.32.4).

Here the dynamic and static aspects of Soma ('thou rushest, seated') are brought together in one of those daring juxtapositions typical of the Ṛgveda. At the physical level, the juice as it speeds forth into the sieve is seen as a deer. It speeds from the womb of the cosmic *ṛta*, but is at the same time seated in the ritual enclosure of the *ṛta* as sacrifice. The two opposites, speed and rest, macrocosmic and microcosmic, plant and god, hint at the quality of omnipresence and the linking up of polarities, spiritual-physical, far and near, up and down. Both at the 'centre' or 'navel' of *ṛta*, whether the cosmic order or its counterpart, the ritual, is meant, and in all manifested spheres – the 'two worlds', heaven and earth – shines Soma, omnipresent and omniscient 'guardian of *ṛta*', like Agni, 'cleaver of the heaven' (*rajasturam*), true 'centre of earth' and 'support of mighty heaven'.[41] In connecting both, Soma is equivalent to Agni who, too, is both head of heaven and centre of earth.

Soma's attribute of spanning heights and plumbing depths also comes out in the following verse:

> In earth's centre and on the mountains has he taken up his abode (IX.82.3b).

As a plant he is found on the mountain and pounded to juice on the Vedic altar, earth's centre. As a god he is present both in the terrestrial and in the spiritual sense, uniting both. The sacrificial enclosure wherein the *ṛta* is re-enacted is always considered the central point of earth which is also Agni's own home, the navel of the world; says the ṛṣi Dīrghatamas:

This altar is the earth's furthest limit;
this sacrificial offering the hub of the universe,
this Soma, the stallion's powerful seed,
this prayer, the speech of highest heaven.[42]

From the 'central point' of the ritual works Soma, 'receiving us in the centre' (IX.10.8). This is not only the sacrificial enclosure as microcosmic centre but also, figuratively, Soma's anchor which is the central point of human consciousness, the point of dissolution and re-solution.

The whole of existence is epitomised in the altar, the meeting-point between the manifest and the unmanifest, the 'navel' of the world. The altar is the final syllable (*anta*) of that ultimate 'beyond' (*para*) which it represents. From the extreme limits to the very core of manifestation, is sacrifice, a constant self-gift for self-perpetuation; the eternally revolving wheel of give and take, of receiving and returning; the sacrificial imprisonment of the spiritual essence in the many forms of material existence and the sublimating or redeeming of this multi-formed matter through fiery purification and sacrificial offering; the eternal balancing, the dynamics of *ṛta*. Everything revolves around the sacrifice which, as the ritual, is the human symbol, the concrete token, of the universal law of give and take, the altar being the miniature representation, the matrix[43] of the all-circumscribing *ṛta* to which all the gods are invited to come.

Prayer, in the priest exalted by Soma, is the very speech of deity, the power of the word, *vāc* manifested in man; through him heaven is made speech in man. The invocation, the word, brahman, in such a state of ecstasy, becomes the key that unlocks the gateway to heaven, the catalyst that galvanises, the dynamo that removes all barriers and reveals the higher states of illumination, the 'secret of the cows' name'.

Soma, giver of illumination, offspring of that supreme consciousness, the transcendent, of that total enlightenment which can only be known partially by the human mind, Soma, the stallion's seed is heaven's revelation in men and gods. Very seed of the transcendent, of life in its spiritual sense, he is that divine urge that compels men to further realisation, to deeper understanding, to fuller bliss, to life more abundant. He, it is, who calls

> the divine races to immortality (IX.108.3).

Through him man enters into a dimension out of all relation to our space-time world, a state of illumination which the ṛṣis called immortality, the converging point to which finally lead all activities in the *ṛta*.

Soma: the tree of life and the sacrificial post

Soma, the celestial catalyst, flows for all creatures

> effecting union and release for our help (IX.84.2c).

The juxtaposition of the two actions, that mean togetherness and separation, which one may consider as opposite, is worthy of special notice. Union with what? Release from what? As usual the text does not give any positive answer. But since this is Soma's action, one may surmise that the 'union' refers to the ṛṣi's entrance into that loftier state of illumined consciousness that enabled him to 'see the gods' and which he called immortality, that state of 'oneness' of which the mystics throughout the ages have attempted to give descriptions. The release effected could be freedom from the thrall of the senses which permits the penetration into a different state of being, Soma's third estate.

This union and release is part of the grand circuit of manifestation, the upward and the downward movement (I.164.47,51) or ascent and descent, the uniting and the separating, the gathering and the dispersing, the offering and the receiving, the dying and the becoming, all expressive of an everlasting rhythm, the pulse of life, the divine sacrifice.

The lord of prayer, Brahmaṇaspati, in his quality of high priest, hence as Agni, 'gathers together and disperses' (II.24.9) what has been offered. Agni, himself described as 'clothed in diverging and converging' forces (I.164.31), energies, or just the 'waters' of space, receives, transmutes and offers man's oblation to the gods who in exchange pour down their blessings both physical (rain, abundance) and spiritual (visions, communion, exaltation, entrance into the divine domains).

The image of the dispersion of the divine life, as Puruṣa/Prājapati, being sacrificed that the many creatures might be, and that of the thousand branched tree of life whose roots, high above, are deeply embedded in heaven (I.24.7) and whose branches come down to our earthly realm, and the reuniting of the divine life offered through the sacrificial post as it links earth and heaven and is itself the symbol of the tree of life, are homologous. The cosmic tree is linked to Puruṣa/Prajāpati and to Soma, called the Sovereign of the wood, or Lord of trees (*vanaspati* in I.91.6), or 'nectar yielding tree' (IX.12.7). The sacrifice of Soma is equivalent to the sacrifice of Puruṣa/Prajāpati:

> For when the gods in the beginning seized (slew) a victim, then as it was drawn upwards, its sacrificial essence flowed downwards, and from it sprang the tree (Śat. Br. 3.8.2.17).

The movement upwards and downwards, the eternal exchange is clearly stated here. Without such exchange there can be no world. Soma, the lord of trees, is killed when the priests press the plant:

> For Soma is a god. But they do kill him when they press him. They kill him with that (stone) (Śat. Br. 3.9.4.17).

This immolation of the plant yields the juice that grants enhancement of life; the plant dies as such that its essence may give life abundant.

The identification of Soma, the celestial nectar, with Prajāpati, the lord of beings, appears in the Ṛgveda, where the bard after summoning nature's artisan, Tvaṣṭṛ, then proceeds to identify both Indra with Soma, called Indu, and Prajāpati with Soma, called Pavamāna:

> Indu is Indra, tawny steer; Pavamāna is Prajāpati (IX.5.9).

As a multifaceted image, the term *vanaspati*, 'sovereign of the forest', in certain contexts (I.142.11) refers to Agni who, hidden in the wood, is its very life, hence its lord, both constructive and destructive, and in other contexts to Soma, as in I.91.6 when Soma is identified with *vanaspati* and is king among plants; but *vanaspati* is also the sacrificial post, the pillar that links earth and heaven, and in itself unites Soma's and Agni's action expressive of the sacrifice.[44]

> Send to the gods, Vanaspati, thyself, the sacrificial draught (I.188.10).

A whole hymn is dedicated to the Lord of the Forest (III.8). Out of the tree of the forest, image of the world tree, man hews and shapes the sacrificial post, his version of the cosmic tree of life:

> In truth, the sacrificer makes himself a ladder and a bridge to reach the celestial world.[45]

Through the tree of life which holds the thousand branched secret (VII.33.9), man partakes of divine knowledge and through sharing in the divine life he integrates himself in it.

The gods scatter the sacrifice 'by means of the sacrificial post',[46] we are told in the Śatapatha Brāhmaṇa, and man gathers the sacrifice and restores it by becoming the sacrifice:

> Man is indeed the first of the sacrificial victims.[47]

This is confirmed in the Ṛgveda:

> The sacrifice is man, providence, our Father (X.100.5c).

The idea of the cosmic tree as the expression or vehicle of the Lord of beings, the active Deity, is similarly attested in the Norse religion where the Yggdrasill is the tree of life whose three roots harbour the world of the dead, of giants, and of men, and also the well of Mimir whose head speaks wisdom. But the word Yggdrasill means the 'horse of Odin':

> It is the means or the vehicle, by which the All-Father descends and ascends, transmigrates, makes his varying avatars.[48]

The *aśvattha* tree, sacred to the Hindus, has also part of its name meaning 'horse' (*aśva*). The tree of life is the manifestation of Deity and as such it is a vehicle, a steed for the Deity. The two images fuse. *Vanaspati* as the cosmic axis is also the means of ascent and descent of men's offerings and of the gods' blessings. But a similar fusion of the idea of 'horse', which in the Ṛgveda is also the great sacrificial victim, symbol of the cosmic life, with that of 'tree' is attested in the Ṛgveda, in an enigmatic verse which refers to Agni as 'the great shadeless leafless steed' (X.24.14). This is confirmed in another verse (VI.13.1) where Agni is compared to a tree whose branches are his blessings to men. Agni, being the 'charioteer of the Transcendent' is both the steed and the tree as well as the essence of the cosmic life that sacrifices all to itself. Hence

> O Agni, sacrifice with thy mouth to thine own body (VI.11.2d).

This could be fruitfully compared to Odin's sacrifice to himself as recorded in the Edda, stanza 139:

> I ween that I hung on the windy tree,
> Hung there for nights full nine;
> With the spear I was wounded; and offered I was
> To Othin, myself to myself,
> On the tree that none may ever know
> What root beneath it runs.[49]

Soma and Agni thus fuse in a most extraordinary way in the image of the lord of trees, on the one hand the king of plants, on the other the leafless steed.

There emerges from these primordial images the great archetype of the tree of life which in giving its own life-sap for the many is constantly dying: a death and a becoming; sacrifice; the law of the universe. In Christian symbolism, the tree becomes the cross of matter upon which is nailed Celestial Man. Vanaspati and Puruṣa meet once again in this age-old symbol of the cross of life and death.

No better summing up of this aspect of the sacrifice can be found than in H. Aguilar's deep study of *The sacrifice in the Ṛgveda*:

> the myth of Puruṣa/Prajāpati reveals to us . . . that the cosmic processes of separation and reunion which constitute the creation and redemption of the world and which are . . . in intimate connection with the structure and internal dynamism of the Lord of Trees, are not merely mechanical, unconscious processes, but they are dominated by a spiritual law of perfect altruism leading to a self immolation, full of understanding and love.[50]

These ideas and images were to be gathered up and fused into the idea of the Supreme:

> The One who rules over every single source
> In whom this whole world comes together and dissolves
> the Lord (Īśāna), the blessing-giver, God adorable;
> by reverencing Him one goes for ever to this peace (Śvet. Up. 4.11).

Soma's third domain: illumination-immortality

The ecstasy of Soma

> opens up, as it were, an unfailing fountain-spring as beverage for man (IX.110.5).

To have drunk Soma seems, as X.119 gives us to understand, to be capable of mighty deeds, creative deeds, extraordinary deeds. In this hymn, the poet, lifted up as though by violent winds, gives vent to his entrancing joy and all conquering dynamism. The consciousness expands to encompass both heaven and earth and rise far beyond, surpassing all known boundaries:

> One of my wings is in the heavens; I have trailed the other below; have I not drunk of Soma? (X.119.11)
> In my vastness I surpassed both heaven and wide earth; have I not drunk of Soma? (X.119.8)

The 'glorious freedom-giving' drops give the bard 'freedom' of space, wide compass, and he asks Soma to inflame him like fire kindled by friction, to illuminate and make him richer (VIII.48.6). The fiery liquid works upon his heart and transforms him into a god, a golden-tongued, sun-eyed, inspired poet.

The ecstasy, the freedom, the unfailing fountain-spring indicate a sphere beyond that of light and sound, the sphere of vision-revelation, inspiration-eloquence, beyond the manifest, wherein the opposites have blended and created that third state of Soma, one characteristic of which is an everlasting renewal from the innermost, hence an unfailing fountain, itself the elixir of *amṛta*. Such a state is described in the Atharvaveda, personifying it as the *ātman*, the innermost self of man which the Upaniṣads were to celebrate as the Inner Ruler immortal:

> Desireless, wise, immortal, self-existent,
> contented with the essence, lacking nothing
> is he. One fears not death who has known him,
> the ātman, serene, ageless, youthful (Athv. X.8.44).

Soma is not only the 'mighty *ṛta*', but the very 'expression of immortality'. The bringing together of the expressions 'cosmic order' and 'state of immortality' in the one deity, Soma, reveals to us a certain hidden significance of *ṛta* that would otherwise be lost to us: the *amṛta*, or immortality,

> secretly hidden away as the supreme domain of the
> sacrificial rite
> originates from the navel of *ṛta*.[51]

The Eternal is rooted in the cosmic order and the cosmic order in the Eternal. The centre of the divine harmony is the timelessness of the Absolute.

Sacrifice is *ṛta* as it establishes the balance between the great inflow and outflow of life, the ingoing and outgoing pulsation of all. The very centre of *ṛta* is the essence of harmony, the point of dissolution of all contrarieties and of resolution; the dimensionless state; that is immortality. Immortality is a state of being out of all relation to our space-time dimension and not the perpetuation of that which by its very nature is limited and conditioned by space and time. The experience of such a state presupposes in the human being a consciousness profoundly widened and deepened in its insights, a possibility to range freely over an all-encompassing because dimensionless reality, an integrity, an integration. Immortality implies the experience of a state of wholeness in which all parts are seen as best fitted to subserve the one end of cosmic order. Wholeness, harmony, oneness, are the hallmark of *ṛta*. In man's case, only in harmonising himself with the whole, in opening his consciousness to a deeper dimension, in integrating himself, in making himself whole and holy, can he realise his essential being (*sat*).[52]

A hidden side to the purpose of the sacrifice can here be glimpsed: a ritual which, in its fullest understanding, would help man to realise the significance of and the reason for integration and the means thereto. In awakening man's inner perception, Soma 'stimulates' the visionary insight into the divine harmony, such action being couched again and again in mythological language. He makes human insight equivalent to Sūrya's gaze, he makes man 'sun-eyed', illumined. So the poets could sing:

> We have drunk Soma; we have become immortal.
> We have gone to the light; we have found the gods (VIII.48.3).
> The vision has been realised into a song. . . . Unto the transcendent one companions came as to a ford. We have succeeded in our song, in praise of prosperity. We have made acquaintance with the immortals (X.31.3).

Soma is the embodiment of the ṛṣis' insights into a realm beyond the physical sphere, a realm marked out by certain potent vibrations that affect the whole human constitution as indeed the whole environment, by luminosity, by freedom of movement, by ecstasy, each presumably a state of consciousness probed by the ṛṣis, culminating in a realm or state of being they called *svar* (heaven), an inner expansion they called immortality, a condition peculiar to the gods, but of which man could partake if he was so inclined, and performed his duties, and integrated himself. Soma is the illuminator through whose means they could envisage the workings of the divine law and indeed were granted the vision of *ṛta*, of cosmic harmony, of truth, of immortality. Only a study in depth of Soma's role in the sacrifice can reveal the full significance which the ṛṣis attached to their ritual.

Agni, the flame divine, the high priest of Order

Passing on to Agni, the high priest of the ritual, we notice that both Soma and Agni are inspirers as well as purifying, dynamic oblation-bearers. To apply the term Agni to the visible fire only is a limitation imposed by our limited human mind. To consider Agni as the epitome of all energies pushes the frontiers of definition to the realms of the cosmic order.

Agni, the messenger of the gods to man, he who dwells at the core of all things and sits in the 'hidden place' of the human heart, the charioteer of the Transcendent, the immortal stallion, is not only 'the arouser of intuitive insights' as is Soma, not only the 'Lord of mind', but he himself is the 'protector of the soma'.[53] The flame-power of Soma is Agni. With Soma's 'heavenly milk' he has been 'anointed by *ṛta*',[54] in accordance with the very law of the universe, he, the 'resourceful inspirer'[55] of men. Both Soma and Agni are said to grant immortality, Agni through his power-of-intelligence (*kratu*), Soma through his exhilaration.[56] Their action precipitates inspiration-exaltation-ecstasy.

Soma is the king; Agni is the high priest, the one that is 'placed foremost' (*purohita*), the 'high priest of Order' who 'celebrates the *ṛta*', the 'cherished one in the rite', the 'god among men in our assembly':[57]

> High priest, bright chariot of the ritual, of every act of worship the resplendent token, thou sharest with every god in might and glory, thou, Agni, guest of men (X.1.5).

As remarked by H. D. Griswold:

> Over against the many priesthoods of men there is the one divine priesthood of Fire, for through Agni alone men worship the gods.

> Agni as the sacrificial element *par excellence* and the archetype of every human priesthood is the very embodiment of ritualistic order.

> Of the three strands of meaning in the rita, 'order', namely cosmic, ethical and ritualistic, the meaning ritualistic order is naturally prominent in connection with Agni.[58]

For such reason he is 'the lord of mighty sacrifice, of all oblations' (VIII.11.4).

His function of 'guardian of the *ṛta*', 'friend, as it were, of mighty order', its 'inspirer',[59] shows him as protector of a rite which, in its reflection of the cosmic order, makes of him the very guardian of that cosmic order.[60] This is his very function, his priesthood. In him the manifest and the unmanifest meet and join in a meaningful fusion. Agni, in fact, watches over *ṛta* in all its aspects and the sacrificial rite is but its symbolic epitome. So he is enjoined to sit

> in the chamber of *ṛta*

> to celebrate the lofty *ṛta* in (his) own domain

> (to) shine forth in thine own home, as guardian of the *ṛta*[61]

cosmic, social, ethical, ritualistic. He is asked, as performer of the sacrifice, to come by the path of *ṛta*, the path mapped out from immemorial times by law:

> With homage and oblation we invoke that performer of the sacrifice by the path of order (I.128.2ab).

For he, as he takes up 'the reins of *ṛta* when sacrificial food is offered' (V.7.3), is the one 'most godlike' and thus capable of

> guiding aright our homage and offering it unto the gods along the path of order (X.70.2cd).

Agni is thus the conveyor of man's aspirations and the giver to man of the gods' grace.

Agni, vehicle of the Transcendent

His sacrificial action, like Soma's, but perhaps in a still more dynamic way, marks him out as the great uniter, he who links the manifest and the unmanifest, heaven and earth, gods and men. Born in *ṛta*, like Soma, he is not exactly said to be the *ṛta*, yet he embodies the vast law in his very function of receiving, assimilating or burning, of giving and returning the essence of the gift, the full circle of life being thereby described. To understand his essential action in the sacrificial rite of linking, purifying, offering and returning is to catch a glimpse of the Vedic meaning of the sacrifice, that which links heaven and earth in one sacred act performed by man and epitomises the whole cosmic process. In such an act all forces work together towards a harmonious end, hence *ṛta*.

> Working 'in accordance with the law' (*ṛtena*), Agni 'reaches out over both worlds' (V.1.7c).

He is 'head of heaven', 'fleet-messenger of earth',[62] the 'deathless oblation-bearer' who, by means of 'vision' (*dhiyā*), 'brings together' (*samṛṇvati*),[63] i.e. he joins vision and offering, man and gods, heaven and earth, ritual and order. He 'holds up', or 'offers our visionary thought'.[64] This function of his to 'unite', like that of Soma, is expressed in various ways, e.g. in the plea to 'worship (with oblation) both wide worlds', or in his ascending and descending; 'he mounts and he descends',[65] he, 'vehicle of the Transcendent',[66] the visible expression of the Invisible. The last is a most important revelation of Agni's function, for as vehicle he steps down the power of the transcendent to within the reach of the worshipper. As vehicle he transmutes man's offering that it may be purified and acceptable to the gods. In encompassing all worlds he is the dynamic charioteer of that which lies beyond these worlds, i.e. the Transcendent. As the intermediary between heaven and earth,[67] as 'conveyor of oblations', as 'ensign of sacrifice', as 'high priest of the gods', the 'inspired high priest, unerring in sacrifices' (V.1.7ab), he enters into the houses of men, he is the 'guest of men',[68] so men themselves have 'established Agni as their priest'.[69]

> In the abode of mortals, the Immortal, the king has sat down, furthering the religious rites (III.1.18ab)

bringing about the means of communication between two worlds.

Agni's function as officiating high priest and mediator between gods and men is constantly brought out by such words as conveyor (*vahni*), charioteer (*rathi*), and the imperative, injunctive or subjunctive forms of the verb 'to worship through sacrifice', √*yaj*. This verb is used to ask Agni to gain for the worshipper the 'favour' or 'protection' (*sumnam*) of the gods Mitra and Varuṇa and the Waters 'in heaven' (*divi*).[70] The word *yoga* which implies the act of joining or harnessing is occasionally used governing *ṛta*: thus 'in the performance (the harnessing) of the *ṛta*' (*ṛtasya yoga*),[71] in the 'yoking' of the rite as a reflection of the universal process, the ends of the cosmic order are subserved, both the performance and the law of which it purports to be the dramatic expression being here brought together in one sacred act, that of the sacrificial ritual.

The stress is constantly made that Agni accomplishes his task among humans according to the dictates of the universal law:

> Established amongst human habitations, Agni, embryo of the waters, achieves (performs his task) in accordance with the *ṛta*.[72]

> Convey our oblation in accordance with the *ṛta*.[73]

True to his inherent being (as indeed are all the gods), he performs that immemorial task of burning, transmuting and offering, of destroying and creating, linking the poles of manifestation through his very being and sacred activity. His very function of offering to the gods man's sacrificial food[74] embodies in itself the law of give and take. In thus consuming man's fuel so that only the essence can reach the gods, he proves himself the 'true sacrificer' (*satyayajam*), 'high priest of heaven and earth',[75] the one archetypal priest on whom all priesthood is modelled.

> Whatever worship we thus perpetually offer to every god, in thee is that oblation presented (II.26.6).

Agni's creative work

Every oblation passes through the flame that burns the dross and keeps but the essence. This creative work of Agni is described in terms of 'human offering', 'making him an axe (*svadhitim*) for the sake of radiance (*tejas*)'.[76] This axe-like sharpening of what is offered may be interpreted as the burning up of the unwanted part of the offering which is mixed up with the essence, leaving the latter only intact, that which is worthy to be received by the gods. This is surely symbolic of the change and transmutation that should take place within man before he can hope to 'enter the *ṛta*', to integrate himself.

Tejas means both sharpness and brilliance, but when applied to human beings (or gods) it refers to that radiance of countenance that manifests as a result of spiritual practices ending in exaltation, so that *tapas* and *tejas* are in a certain sense related, one emphasising the process, the other the manifestation of that which, at the physical level, is seen as radiance, a shining from within without. Those who are 'sun-eyed' (*sūracakṣasaḥ*), i.e. radiant and are therefore wise, with Agni as their tongue, 'flame-tongued' (*agnijihvaḥ*),[77] 'foster the *ṛta*' and are fostered by it.[78] They grow in understanding of, and through adherence to, the Order.[79] These are the gods. But men can follow in their footsteps.

This should throw some light on the meaning of that oft-mentioned radiance or splendour which characterises Agni as not merely the shining flame, but the illumination by means of which the deity is credited with an understanding of the cosmic order and which grants that very understanding to those who have kindled the flame within themselves; the flame is but the outer token of the inner state. Agni is not only the flame that burns the offering, but he who inspires or kindles man's intelligence, the resourceful inspirer or 'enlightened knower' (*kavikratuḥ*), who brings about this spiritual exaltation in man, which exaltation is personified in Soma. Aroused through the flame and exalted by Soma, man through the action of both deities, enters into the divine realm:

> through the power of his resplendence he clasps the rein of *ṛta* (V.7.6).

By means of Agni man harnesses the ritual for purposes of reflecting the cosmic order and coming to terms with both the vision of it and the living in accordance with it. This is Agni's task, as the cosmic priest, and by reflection, it is that of the human priest. Noteworthy is the specific mention that this clasping is accomplished through the 'power of resplendence', that illuminating knowledge which allows the god to perform his function, and the worshipper to enter into the spirit of it and understand what lies at the back of it.

The insistence upon the understanding, insight and wisdom which Agni incarnates and which he imparts to man is an indication of the ṛṣis' own stand for knowledge; the sacrificial ritual is not a mere mechanical procedure, but it is fraught with meaning, and in its course the great high priest himself bestows upon the performer his own wisdom:

> For thou of gracious intelligence and unerring insight hast been the charioteer of lofty *ṛta*.[80]

As 'wisest of priests'[81] he, with 'wise insight',

> attends in every house to the sacrificial rite (I.128.4ab).

Agni, the knower, the releaser

So on him 'the very wise',[82] men are enjoined to 'meditate', for he is the 'knower' (*cikitvām*), the 'knower of all wisdom',[83] the 'knower of generations', the 'all-knowing one'.[84]

No wonder that to him goes that most ancient of human cries:

> Shine forth auspiciously upon us that we may reach wide understanding (VII.3.10).

Agni's action on man as 'the finder of light'[85] is similar to Soma's:

> Thou prevailest over every thought;
> thou promotest the intuition of the wise.[86]

> May he stimulate us to vigour and vision.[87]

> From thee O Agni inspired-wisdom, from thee intuitions, from thee accomplished utterances are generated.[88]

His, like Soma's, is the quality of the sage: intuitive insight (*manīṣā*), wisdom imparted through inspiration, the poet's gift, and as a result, wise utterances (*ukthā*). Like Soma he is a ṛṣi.[89] To him goes the same plea as to Soma to 'open out' the intuitive insight, hence to release the inner vision and offer it in his function of high priest.[90]

To Agni's ministry is ascribed a very subtle role, that of inner illumination, and this quite apart from any purifying and ritualistic role that he assumes. His action is to

> bring to birth within the heart a thought as light (III.26.8ab)

causing the inspired thought to come to consciousness which is then shaped into a song:

> he purifies the flashing song with threefold filter (ibid.).

The thought becomes a colourful song since *arka* means both a ray, flash of lightning and a hymn or praise. Born in the heart, seen as light, expressed as song, with Agni as both its arouser and its purifier, it is then offered to the devotee's god, whoever this might be:

> thought (*matis*) moving forth from the heart fashioned in praise goes to Indra, the Lord (III.39.1).
> With pleasing hearts let us approach-with-veneration, Agni the pleasing, beloved of many, the sharp-rayed purifying flame (VIII.43.31).

The kindling of Agni is thus not simply the physical action but as well the mental act (of which the physical is an extension), the inner process of awakening the flame of aspiration-inspiration-ecstasy, the result of which seems to have been that vision (*dhīḥ*) which translated itself into a hymn, which granted the ṛṣi a penetrative insight into the laws of the universe, an intuitive perception which is itself offered up to the gods in the sacrifice[91] to be renewed and received again.

For Agni is not merely the fire on the altar or in the wood, but

> the one ocean, the foundation of riches, of many births (that) shines forth from our heart (X.5.1).

Here is set forth unequivocably the intimate connection between Agni and man. This is further confirmed in the following:

> Seated in secret place . . . men find him there whilst meditating and reciting their mantras which they fashion in their heart (I.67.2).

The 'guest' in 'every house'[92] is not only the fire on every altar but the flame in every heart.

As in the case of Soma we should also observe that the *yajña* was not conceived merely as a ritualistic process to be meticulously performed, but, in a deeper sense, it was considered in terms of an offering of self welling forth from the 'heart'. This is clear from the following verse where the offering (*havis*) is specifically mentioned as shaped in the heart:

> Agni, to thee we bring with our song this offering shaped in our heart.

The second part of the verse is even more revealing:

> Let these be oxen, bulls, cattle to thee (VI.16.47).

Here is openly claimed that the true offering fashioned in the heart is worth all the cattle and is evidently to Agni what wealth in kine is to man. In Ṛgv. X.91.14 a similar thought seems to be voiced, for there also horses, bulls, cattle, all the valuables of the Vedic people are mentioned as offered up to Agni as the sacrificer brings forth

> the cherished intuitive perception from his heart.[93]

The keyword in these verses is surely the 'heart'. Whatever is most treasured springs from the heart to be offered up in the fire and that is likened to whatever is most prized in everyday life, the essential thought being the oblation as a heartfelt offering.[94]

Conclusion (1)

The action of both Soma and Agni upon the worshippers in the performances of their function is thus revealing of the fundamental meaning of the rite: man's epitomising of, and participating in, the cosmic 'sacred work'. One may draw the conclusion that just as *ṛta,* in its ideal sense, has to be dynamically actualised on earth by godly as well as human agency, both social and ritual, so the enlightened seer, striving after full illumination, has to reconcile the inner vision he perceives in his highest state of exaltation with the outer, everyday reality of hardship and conflicts and finally to experience and live in that state of enlightenment which is pure harmony and transcendence of the pygmy self, a state which the ṛṣis qualified as immortality.

The intimate connection between Soma and Agni's action can be perceived in the state of consciousness which Agni grants through his wise insight, a state which is akin to

Soma's enlightenment and by means of which the gods reached immortality.[95] Agni, 'lord of abundant *amṛta*', himself is made 'centre of immortality' by the gods[96] as he is both their flame of aspiration and their very achievement. They protect the 'boon-bestowing' (*draviṇodām*) Agni as their own immortal state.

Two stanzas, each concerning Agni's and Soma's strengthening influence, when juxtaposed, reveal the effort needed to attain the state of immortality:

> Agni, the knower, procures heroic-might,
> procures vigour, enhancing it for immortality (III.25.2ab).
> 'Far-seeing mind, enlightening, light-bestower . . . the
> mighty one (Soma) strives after his third estate (IX.96.18)

that third estate which seems to be the 'supreme domain of the sacrifice',[97] its innermost secret, immortality. In the one case, the heroic power is enhanced[98] through Agni's action, in the other, the desiderative form of the verb 'to wish to gain' (√*san*) shows a striving after. This striving might be indicative of a transmutation process that operates in man at a certain stage of enlightenment, a transmutation that permits him to force entrance into another level of awareness. In both cases effort is implied and in both the deities seem to become the very embodiment of human aspiration, enlightenment and exaltation.

In the same order of thought, Soma, 'lord of valour', giver of 'vigour' (*vayodhāḥ*) is asked for a share of his riches, those riches that are mental ability and insight.

> Bring us that resplendent courage that will find heaven
> (IX.106.4)

is a plea addressed to Soma, but could equally be addressed to Agni.[99] Only those men who are purified and brave, only those fire-proved, heroic souls can gain the boon of immortality. The power envisaged seems to be rather that which storms the kingdom of heaven than that which destroys man-made forts. *Ṛta* can only be apprehended, and its very core, immortality, can only be attained, by the pure, by the tried, by heroes.

The heart, crux of the sacrifice

The question of the heart is really the pivot on which centres the depth of the significance of Vedic sacrifice, a significance so totally missed by early Western exegesis. The values of the heart are stressed either directly or indirectly as when the poet makes his offering from the heart or hopes it will touch his god's heart. Its role in the fostering of vision and the fashioning of songs has been considered with regard to Agni, the kindler of inspiration.

Of the several examples of Agni's great role in human life, the point to be emphasised here is his being considered:

> the one ocean, the foundation of riches . . . (that) shines forth from our heart (X.5.1).

He sits 'in secret place' (*guhā niṣīdan*), i.e. in the human heart

> where men find him whilst meditating and reciting their mantras which they fashion in their heart (I.67.2).

The flame in the human heart is thus the sacrificer, the outer fire being but the physical token of an inner dynamic, all-kindling energy that sets all things on their onward course, and in man directs him onto the path of immortality.[100] This has been disregarded in the evaluation of the Ṛgvedic sacrifice. The human heart is the seat of the offering, the enclosure wherein is enacted the drama of give and take, the eternal exchange, the divine transmutation whereby the lower is uplifted and in the process may suffer by losing itself, and the higher is exalted and in the process there is effected a union whereby the joy of oneness is fully experienced.

Further examples should clarify the question of the heart's role. The offering of the praise, song or hymn, the end-product of the vision is, as we have seen, part of the sacrificial oblation, and perhaps dearest to the heart of the worshipper because it was his own effort, expression, creation. In this respect the following verses reveal a great deal as to his beliefs, hopes and feelings:

> On whose heart shall we clasp this divine paean rich in oblation, most beloved among the immortals? (IV.43.1cd)

> My songs-of-praise, high-aspiring, speed-forth as messengers to Indra . . . uttered by my spirit to touch his heart (X.47.7abc).

> May this song-of-praise be foremost in touching thy heart, in blessedness (I.16.7ab).

> O Maruts, ye gods, to you with homage is offered this song-of-praise shaped with heart and mind (I.171.2ab).

Soma's boons of insight and intelligence touch the heart (*hṛdispṛśas*) (X.25.2a). To touch Varuṇa's spirit and win his mercy the poet attempts to 'bind' him with his songs (I.25.3).

Brahman companions, as already mentioned, come together for worship 'with intuitive flashes fashioned in the heart' (X.71.8), whilst others who pose as Brahmans wander elsewhere. By means of insights gained within the heart the Vasiṣṭhas 'secretly' approached 'the thousand branched (tree of knowledge?)' (VII.33.9ab) and penetrated to the realm of Yama, the king of the kingdom of death. Pondering in their heart the sages discovered the connection between the *sat* and the *asat* (X.129.4).

All these examples point to the heart as being the recognised centre of both receiving and giving, of inner perception and understanding, in other words, of a spiritual insight far removed from the rational faculty. This is further enhanced in a hymn to the All-gods (Viśvedevaḥ) where Varuṇa 'reveals' or

> opens out his thought by means of the heart (I.105.15c)

as a result of which a new manifestation of *ṛta* is made apparent. The word 'heart' must therefore have been some kind of technical term that comprised the meaning of understanding-intuition-wisdom or spiritual insight. Revelation comes through the heart as the seat of synthesis. It is in the heart that Varuṇa placed the power of intelligence[101] which combines awareness, understanding and intuitive perception and yields the higher vision. This is the meaning of the later word, *buddhi*.

The heart[102] is thus the secret or hidden place (*guhā*) wherein knowledge, the divine treasure, is stored. Several verses refer to that knowledge, received in secret (*ninyam*).[103] The language of some of these is somewhat recondite: the Vedic poets had evidently their own technical terms to refer to certain states of consciousness and perception beyond the normal, mundane knowledge and awareness:

> Agni the knower has revealed this inspiration to me as
> the hidden footprint of the cow (IV.5.3cd).
> That which was established as Viṣṇu's supreme seat,
> therewith thou guardest the cows' secret name
> (V.3.3cd).

The name being the essence of a thing, the cows' secret name would refer to the deepest mystery of nature which can only be apprehended in 'heaven' or in the state of ecstasy granted by Soma.

That these revelations are brought about by Agni or Soma, the ministers of the ritual, is another pointer to the specific function of these rites. The sacrifice is a prayer and prayer an action, a communication, hence sacrifice is sacred work (*apas*). The shaping of prayer and the offering of the oblation, or of the vision-inspired-thought, or of the recitation, all have their roots in the heart. The Lord of prayer, Bṛhaspati, is worshipped

> with this new and mighty song (II.24.1b).

> To thee O Agni I bring the offering with the sacred utterance (*ṛk*) fashioned in the heart (VI.16.47ab).

To the Maruts is offered the praise 'shaped in the heart and mind', and, to Indra, the Soma drops are 'offered from the heart'.[104] The original oblation offered by man was Manu's, the first human, by means of his spirit, with kindled fire.[105] That Agni

> found the path to prayer (VII.13.3c)

would imply that sacrificial offering is the means of communication with the gods, hence the path of invocation which is prayer. Such heartfelt offering is what brings reward whether as protection or as boon, or as the mere presence of the deity which itself is a blessing:

> Whoso with oblation, with god-directed spirit and with chrism, pays homage to ye, O Agni-Soma, protect his way, guard him from distress (I.93.8abc).

'Strengthened by prayer'[106] both Agni and Soma grant the worshipper ample room (*uruṃ lokam*) or free scope for his sacrifice.[107] The response of the god seems to depend upon the efficacy or the power of the invocation (*brahman*). This is an important factor in the bringing together of gods and men. When there is no heartfelt longing or call for communion, there is no power, hence no response, no revelation, no nearness or divine presence.

The sacrifice (*yajña*) that is performed 'cut off from prayer'[108] does not please Indra. Likewise the Soma juice pressed without prayer (*abrahmaṇo*) finds no favour in his eyes.[109] For Indra thrives not only on Soma but on prayer as well as the sacrificial oblation.[110]

The source of the inspiration is thus rooted deep in the human being, in that central focus whence inspirations well forth and insights into the world processes emerge. The poets describe themselves as 'devout' and 'god-serving',[111] whilst seeking Agni, their 'protector', both father and mother.[112] The states of grace and piety are thus a prerequisite to their finding the god and actively participating in the oblation.

All this shows prayer not only in its sense of powerful invocation but also of heartfelt longing, as an essential element of the sacrificial offering. When we realise that brahman is not prayer in our ordinary sense of the term – a mere asking or even simple attuning to a higher power – but a dynamic, god-given (*devattam*) power that man can use to invoke the noumenal or come into the gods' presence, then the sacrifice, such as it was conceived in Ṛgvedic times, can no longer be regarded as a mere bargain or a 'hocus-pocus', but a ritual dramatising man's inner transformation – his appeal, his commitment and participation – an action that brings him in harmony with the universal order, a ritual integration.

To quote from R. Panikkar: 'the common underlying assumption or implicit belief – is that in the act of prayer Man is sharing in the central dynamism of reality and penetrating into the heart of the world. Prayer is truly Brahman.'[113] Prayer, offering, sacrifice, are essentially sacred acts of the human heart. They are steps indicative of man's integration into the cosmic harmony.

Conclusion (2)

The detailed examination of all the verses surveyed in this chapter should serve to show the vastness as well as the subtlety of the conception that lay behind the Vedic ritual and the superfluity of those Western critics who dismissed it as meaningless twaddle. Very subtle are many of the hints whose depth of thought hinges upon the understanding of a few keywords which have all too often been bypassed. As A. Danielou well understood: 'The fire sacrifice became the essential instrument of man's participation in the cosmic sacrifice. . . . The Vedic ritual aimed at resembling more and more perfectly the very ritual through which the universe exists.'[114]

Purification, aspiration, recitation, visioning, communion, togetherness, are expressions of the stirrings of the soul, its strivings, its faith, hope and trust; all these are fostered by the Vedic sacrificial ritual. They are further pointers to man's attempt at integrating himself in the vast order, of being himself the living embodiment of that harmony which means true life. The *yajña* epitomises the action of give and take, which itself is a shorthand symbol of the law of transformation and of the vast rhythm of the universe. In participating ritualistically in this constant exchange, man learnt to integrate himself in the cosmic order. This could only be done through the inner purifying of himself and the coming to terms with society of which the purificatory rites and the assembling together were the outer mark. The ritual forms a complement to the norms of social life as established by Vedic ethics. It galvanises these norms into a purposeful reflection of the cosmic process and gives man a condensed view of the meaning of it all.

The Vedic point of view, such as it may be reconstructed from the Ṛgveda Saṃhitā, is best summarised by R. Panikkar:

> Vedic Man is fundamentally a celebrating Man, but he does not celebrate his own victories or even a nature festival in company with his fellowmen; rather, he concelebrates with the whole universe, taking his place in the cosmic sacrifice in which all the Gods are engaged together.
>
> Ṛta is . . . the actual functioning or rather the proper rhythm of the sacrifice, while sacrifice is that which causes things to be what they are. By sacrifice Gods and Men collaborate, not only among themselves but also for the maintenance and very existence of the universe. Reality subsists, thanks to sacrifice. But this truly primordial sacrifice is not left to the whim of either Men or Gods; it has an internal structure and mode of operation, namely, ṛta.[115]

So long as the ṛṣis were inspired to compose their hymns, so long did the ritual remain a live, meaningful symbol and vital action, the outer token of their understanding, their vision, their livingness. But when the 'spirit' of the sacrificial ritual was almost forgotten and priests only clung to the 'letter', these rites still served, for many centuries, the purpose of bringing people together in worship, hence in some kind of harmony and in the remembrance of an originally sacred action. *Ṛta* was still somehow mirrored in *yajña*.

Part V
COSMIC HARMONY

FOURTEEN

Ṛta, world-embracing power

The question whether *ṛta* held, as H. Lefever[1] claims for it, the status of a 'world-embracing power' in the sense of being considered a deity in its own right, cannot be entirely satisfactorily solved. The pros and cons seem to weigh about equally in the balance.

As a unity and the one universal order, *ṛta* does encompass, as we have seen, each of the three norms of established order, the natural, the ethico-social and the religio-sacrificial. By its power Varuṇa and Mitra govern all the world (V.63.7). H. W. Wallis claims that 'The rita was . . . in itself too stationary, too conservative, and purely regulative to be endowed with the individual life necessary to a god; it remained an abstract idea and was not an object of direct adoration.'[2] A similar opinion is voiced by A. B. Keith:

> As a deity . . . the Ṛta does not obtain an established rank. The occurrence of such phrases as the idea of the Dawns coming forth from the place of Ṛta, or of the place of sacrifice as the seat of the Ṛta, or of the charioteer or the wheel or the steeds or the vehicle or the ship of the Ṛta, shows nothing more than the natural concreteness of expression of the Vedic age; there is no prayer to it, and in all the mass of deities invoked in the cult the Ṛta is not included.[3]

This is very far from accurate. Injunctions to meditate on *ṛta* can be found as well as invitations to pay homage to *ṛta* and straightforward invocations.[4] Direct worship may not have been as freely lavished upon it, as being the universal law, hence impersonal, as it was upon the more personal gods that both serve it and rule by its means and whom the worshippers could feel as being closer to themselves. Nevertheless certain verses seem to indicate that it could have been invoked as a deity and, if so, other passages where it is mentioned together with some of the devas may also imply an invocation which so far has escaped notice.

Wallis's definition of what is necessary for a god to qualify as such, namely 'individual life', shows the relativity and limitation of godhood, since individuality is a limitation. The Vedic gods, just as the God of any monotheistic faith, are bound by the very 'individuality' with which they are endowed by their devotees. Only, the Vedic mind went further and conceived of the one transcendental law as the only fit expression of the Absolute which itself stood beyond human speculation.[5]

The Vedic conception of divine power

The whole issue centres around the Vedic conception of divine power. Certain categories of powers are evident – if we may thus subdivide these – in the Vedas:

1. those great cosmic powers or devas who are both impersonal and personal, some of whom appear both as powers and as persons (Dakṣa, Aditi, Agni, Bṛhaspati, Soma), and others with a very marked personal character (Varuṇa, Indra, the Aśvins);
2. such seemingly impersonal powers as *ṛta*, *uttānapad*, *iḷā*, *tapas*, *tejas*, *sahas*;
3. and personifications of abstract nouns, such as *śraddhā*, faith, *manyu*, zeal, mettle, fury, *takman*, fever.

Three hymns, e.g. one to *manyu* (Ṛgv. X.84.4), one to *ṛtu* (season) (Ṛgv. I.15), one to *śraddhā* (faith) Ṛgv. X.151), show that abstract conceptions or qualities such as faith, personalised into an independent power capable of being invoked, could be addressed with a view to filling the worshippers with their essence:

> We call upon Faith at dawn; we call upon Faith at midday;
> we call upon Faith at the setting of the sun;
> O Faith inspire us herewith with confidence.[6]

This is what Keith describes as 'deifying',[7] which seems rather to be an exaggerated claim.

Power in the Vedas always has an element of intelligence which, in human thinking, endows it with a certain amount of personality. As the supreme law, *ṛta* is an impersonal power, the underlying regulator of all phenomena, whether at the natural level or at the level of its reflection, the human. At the same time it has a certain quality which the human mind apprehends in terms of the personal. The very fact that the Vedic poet could address *ṛta* thus:

> I solicit the *ṛta* controlled by the *ṛta* (IV.3.9),

shows man invoking the vast cosmic order itself by means of the rite. He may have represented it to himself as an almost personal entity whose attention he is requesting. To 'solicit, entreat, request, glorify' ($\sqrt{}$*īḍ*) can only be used of intelligences whose attention is believed capable of being turned to one's self. On the other hand, he may have also just expressed his admiration ('I glorify' can also be a translation) in a superb gesture of both praise and entreaty. *Ṛta* is both impersonal and personal, like all other powers in the Vedas. Personification in the Ṛgveda is never clear-cut. The great cosmic powers themselves, the devas, are beyond personification, and yet not completely impersonal. To the Western mind they have not yet 'acquired' a full personality as they do in post-Vedic mythology and in such late mythologies as the Greek and the Roman. The human trend has been towards anthropomorphising the numinous and thus making it more accessible to average human intelligence. It is no sign of advance in or development of understanding. The Vedic cosmos is ruled by intelligent Law which manifests through the action of intelligent units, its dynamic agents, who all one-minded, one-intentioned, unerringly proceed to the one purposeful accomplishment (VI.9.5).

Ṛta, universal law

Ṛta is, as we have seen, the supreme law from which all other laws are derived, the order of the universe, the blueprint of what, by means of its actualisation in the phenomenal world, will bring about the manifestation of the Transcendental through the co-operation of men and devas and therefore the first to be conceived at the origin of the universe.[8] The gods follow its statutes (*vrata*), hence are its charioteers (*rathi*). Soma flows under its *dharma*, its commandments (*pradiśaḥ*) make Indra mighty.[9] To it the gods are faithful (*ṛtavari*), in it they are born (*ṛtajāta*). Scholars (e.g. A. Bergaigne, H. Lefever) point out that *dharma*, *dhāman* and *vrata*, which are the three main law terms apart from *ṛta* (already considered in connection with the gods' action), govern the *ṛta* in the genitive, but *ṛta* never governs any one of them.[10] Many other expressions[11] are used as qualifying *ṛta* in a way in which *satya* or *yajña* are not used in the Ṛgveda. *Ṛta* in such cases could be considered as taking on the attributes of a deity. On the other hand it is observed that *ṛta* is also used in subordination to some of the gods, e.g. Varuṇa and Mitra.[12] This apparent paradox is discussed by H. Lefever who explains that the gods act by the 'power' of *ṛta* and only in this sense does it seem to be subordinate to them. He points out that

> Ṛta as the actual order realised in the world, is a product of the activity of Gods and men, an activity directed in accordance with the transcendent ṛta. This is the significance of such expressions as 'By ṛta upholding ṛta' (1.23.5 – cf. V.15.2).
>
> Ṛta . . . is transcendent as the governing principle of all life and activity in the sphere of space, time and will. To this sphere belong both gods and men. . . . Ṛta is, on the

> one hand, the law of their own nature and so is immanent, while, on the other hand, it is the law which they must obey and yet are free to disobey, and so is transcendent.[13]

The harmonious movement of all things shows the ordered course, the transcendent *ṛta* as it is reflected at the phenomenal level. But the actual operation of this harmonious movement in the space-time world is the task of the gods who 'harness' the power of *ṛta*. This power is described as 'efficient' (*dakṣam*) and coming 'from lofty heaven' (*divo bṛhato*). This is *ṛta* which the gods (Mitra-Varuṇa) yoke (*yuñjāthe*) to the work (*apas*) as to the pole is bound the ox. Hence the gods are both its charioteers (*rathī*) and its guardians (*gopāḥ*) or superintendents. They 'proclaim *ṛta*' i.e. their every action is a manifestation of *ṛta* because they only work in accordance with it. So Indra 'harnesses' or 'steers' the *ṛta*, Viṣṇu gives the worshipper his 'share of *ṛta*', i.e. maps out for man what befits him. Of all the gods the Adityas are the most closely connected with the supervision of the working of *ṛta* both in the phenomenal and in the human fields. Hence the '*ṛta* of Varuṇa', the '*ṛta* of Mitra', refer to that sphere of *ṛta* which is under the superintendence of these gods and do not necessarily mean subordination of the *ṛta* to these gods.[14] Hence, by Mitra-Varuṇa's

> *ṛta* is *ṛta* firmly established (V.62.1).

However much the gods may use the power of *ṛta*, i.e. act in accordance with the law, they ever remain its custodians (*gopāḥ*). To be 'born of *ṛta*' may have been considered the highest praise. Occasionally we find the culmination of praise extending as far as identification with *ṛta*, but this may be the highest blessing that the poet may imagine for his deity. Thus Soma is 'the vast *ṛta*'. Thus Mitra Varuṇa, mighty lords, 'are the great *ṛta*'.[15]

These few examples serve to show the extreme reverence in which the *ṛta* was held by the ṛṣis. This could be a pointer to their having considered it as an impersonal deity much in the same way that Brahman was held in the time of the

Upaniṣads. Only *ṛta* is the way to the source, Brahman is the source.

Ṛta as deity

V. M. Apte brings forward some quotations which, in his opinion, show that *ṛta* ranked as a deity in the Vedic mind. He writes: 'The mighty (*mahad*) ṛta figures as an independent deity among deities like Aditi, Dyāvāpṛthivī, Indra, Viṣṇu and Savitṛ invoked for protection in X.66.4.'[16] This verse, literally translated, runs thus:

> Aditi, Heaven and Earth, Ṛta the mighty, Indra, Viṣṇu, the Maruts, the lofty empyrean we call upon the gods, the Adityas for help, the Vasus, the Rudras, Savitṛ of mighty accomplishment.

Here *ṛta* is certainly on a par with the other gods and one could affirm that if Aditi is considered a goddess in her own right so in this passage *ṛta* could also have been called upon as a deity in its own right. But the term *svar bṛhat* (lofty empyrean) should also be noticed. Is this different from *dyaus*, Heaven, and if so, is it considered a deity? Not necessarily, in which case *ṛtaṃ mahad* may not be a deity in the accepted sense of the term. This does not prevent it from being called upon together with the recognised deities.

V. M. Apte then quotes a somewhat controversial verse:

> The soma juice is pressed for Mitra and Varuṇa to drink, the pleasant juice for Ṛta to drink.[17]

The word *ṛta* here again seems to rank as equal in status with the gods quoted alongside with it. The whole argument of the status of *ṛta* hinges around the idea that personal gods are considered (by Western scholars) higher than an impersonal power. That may not have been so in Vedic times and there is no reason why the 'personal' should rank higher than the 'impersonal'. The only difference, to man, between personal and impersonal powers being that man, because of his nature, will seek protection, refuge, comfort, etc. from a personal rather than an impersonal power, endowing the former with what he considers his own higher attributes whilst the latter remains more or less abstruse. It may be noted that in the Upaniṣads the impersonal Brahman assumes precedence over every personal deity. E. W. Hopkins considers that *ṛta* was personified 'as a divine personal Power' towards the very end of the Ṛgveda but that

> in no passage is Varuṇa subject to Rita as a superior power, but rather Rita (Right Order) has its very source in the wise spirit (Varuṇa) and all the laws of Varuṇa and of the lesser gods are expressions in concrete detail of the divine Order which emanates from heaven, born . . . of the religious fervor of the Creator.[18]

There is however no extant verse stating that *ṛta* has its source in the wise spirit, Varuṇa,[19] unless Varuṇa be equated with the Ordainer (*dhātṛ*) of Ṛgv. X.190.3, and even here *ṛta* seems to antedate the creator, or Time the great fashioner. It is evident that Western scholars find it difficult to conceive of an impersonal power as a fit object of worship.

Ṛta in the above-quoted verse (Ṛgv. 1.137.2) is used in the same dative case as Mitra, Varuṇa. The construction of the sentence also shows this kind of equality: *pītaye* (to drink) applies to Mitra-Varuṇa and the second *pītaye* applies to *ṛtāya* (for *ṛta* to drink) in strict parallel to the first part of the sentence. R. T. H. Griffith translates the phrase thus:

> Juice waits for Mitra and for Varuṇa to drink, fair juice for drink, for sacrifice.[20]

thus giving a different twist to the meaning, and avoiding what to the Western mind seems inconceivable, namely that an impersonal power be raised to godly status. The verse may, on the other hand, simply mean that the offering of the Soma to *ṛta* is an act of obeisance to the cosmic order, a symbol of man's participation in its process.

Another Ṛgvedic verse states:

> Pay homage for us to Mitra Varuṇa, pay homage to the gods,

> O Agni, pay homage to the mighty *ṛta*, thine own domain (I.75.5).

According to A. Bergaigne[21] the mighty *ṛta* is here the sacrifice that is offered to the gods, which would make the verse read 'sacrifice for us a mighty sacrifice' or 'offer up the mighty order'.[22]

Whether personified or not, *ṛta* remains that transcendental harmony, that Cosmic Order to which man aspires and in so doing endeavours to regulate his life in accordance with principles which he reads in it, establishing the norms of social and ethical harmony which the ritual is meant to mirror. In the words of A. C. Bose, 'the order observed at the ritual is a symbol of eternal Order, including the eternal statutes of moral life'.[23] Perhaps *ṛta* was the best expression of Deity in manifestation that the ṛṣis could conceive – the vast course of life that flows through the multitudinous, separate existences and phenomena, the variegated beauties of the universe. Homage could be paid to it, during the sacrificial rite, as the very reflection in the cosmos of the impersonal Absolute, *tat*. Hence the praise:

> Firmly based are the foundations of *ṛta*, shining in beauty, manifold are its beauteous forms (IV.23.9ab).

That, in the final analysis, all creation pays homage to the *ṛta* is stated in bold imagery:

> To Law wide Earth and deep (Heaven) –
> supreme milch-kine – render their milk (IV.23.10d).

Every act performed in conformity with one's nature is a homage paid to that Supreme Law from which springs the very law of our being. In this sense also can be understood the universal homage (or sacrifice) paid by the gods and all creation to the One Law which is the very essence of the divine life – the Puruṣa sacrificed at the origin of all.

Ṛta, subject for meditation

If the world order is fit for admiration, it is also fit for meditation. With this thought in mind we could glance at certain hints given out in Ṛgv. IV.23.8–10. Verse eight begins with the realisation that

> Ancient are the invigorating-draughts of *ṛta*
> Contemplation of *ṛta* removes transgressions.

The word *śurudhaḥ* meaning 'riches', 'comforts' or 'invigorating-draughts', shows *ṛta* as pouring out blessings that have been of old, therefore that are ever present, that bring one to a state of holiness, since an end is put to sin. The counsel given here is similar to all mystical teachings which advise concentration upon a chosen deity by which means the devotee will acquire the attributes of that divinity. The stanza ends with the comment that the song-of-praise (*śloka*) of *ṛta* opens the deaf ears of the priest; therefore to praise the *ṛta* is to effect some transformation in the human being. Verse ten emphasises a similar thought:

> Whoso harnesses the *ṛta* wins the *ṛta,*

in other words, who obeys nature masters nature because he has learned to use her law in her own proper way. Nature here can also be considered as epitomised in the ritual. Thence does

> the might of *ṛta* fervently speed onwards,

in this way is all accomplished in accordance with the *ṛta*.

These verses provide an example of the possibility of a personal kind of relationship between man and that impersonal power behind and within the universe, and the potent effect the latter can have on the former. The same type of prayer is observed in the following:

> Pay attention to the *ṛta,* O attentive one, to the *ṛta,* split open the many streams of *ṛta,*[24]

in other words, open the sources of *ṛta*. The Vedic seers were on the alert to catch glimpses of the Cosmic Order in their vision and performed all works in accordance with the 'summons of *ṛta*' (*ṛtasya preṣā*), with the 'visionary insights of *ṛta*'. This concentration on the *ṛta* is like an opening of one's mind to the divine inflow. The direct action of the *ṛta* on the human being is implied as well as its beneficent effects. It was thus considered a universal, beneficent power to which man could turn. *Ṛta* is the way to the Transcendent. It is the only fit expression of the nature of the Transcendent, as mirrored at the manifested level, of that which the Upaniṣads called Brahman. Cosmic harmony images the Unmanifest and is therefore a worthy object of meditation. It is the language of the Unmanifest.

Ṛta is both a differentiated and an integrated Power that yet stands beyond differentiation and integration because it makes both actions possible. Sanctity conferred upon *ṛta*, or making it a deity, is man's ultimate expression of acknowledgment and reverence offered to something that he senses as far beyond himself and too superior to be reached even by imagination. His worship of *ṛta* is indirectly manifested when he lives in accordance with the highest norms that he is capable of conceiving and when he pays homage through thought, word and deed to the gods who are the custodians of *ṛta*. To attune oneself to *ṛta* is to take the right direction to the right goal. It is the return to the source. It is integration. Hence 'from the navel of *ṛta* is born immortality'.[25] This is the essence of the Ṛgvedic vision and its message. This is the Vedic revelation (*śruti*).

FIFTEEN

Conclusion: *Ṛta*, cosmic harmony

The whole trend of this investigation into the significance of *ṛta* has been an endeavour to examine the many facets of the Ṛgvedic vision of *ṛta*, of a cosmic order in which all parts throb with dynamic life and play an effective role subservient to the whole. One cannot assert that *ṛta* is just the fixed order of things, or just truth, or just law, but its meaning includes all these and goes far beyond, for *ṛta* is transcendental; it stands for universal truth, universal harmony, universal law, universal order, universal life; in standing for all these conceptions which can only be envisaged in man's loftiest flights, it points to the richness of its underlying meaning; it is the manifested expression of the Transcendent.

The school of thought based upon nineteenth and early twentieth century conclusions which still influences scholars nowadays[1] as well as the more popular types of expositions on India's earlier forms of religion, and which considers the Ṛgveda as a collection of more or less primitive songs, of nature worship, the tribal cult of simple minds, a 'naive religion', needs complete reviewing. S. Radhakrishnan summed up the general position in the following words:

> A study of the hymns of the Ṛgveda is indispensable for any adequate account of Indian thought. Whatever we may think of them, half-formed myths or crude allegories, obscure gropings or immature compositions, still they are the source of the later practices and philosophies of

> the Indo-Aryans, and a study of them is necessary for a proper understanding of subsequent thought.[2]

Fortunately the latest trend is against such an ill-considered, somewhat superficial and biased approach. With the work of J. Gonda, L. Renou, R. Panikkar, K. Werner and others, the Ṛgveda is being studied in depth with an open mind and the claim of the ṛṣis to spiritual insights is being investigated and more and more vindicated.

That the Vedic hymns show a religion deeply imbued with the poetry of and the feeling for nature, and man as essentially part of nature, is all too evident. Vedic man lived in such proximity to nature that there could be no divorce in his thought between the latter and himself. For him natural phenomena were as divine as his own divinely inspired and most treasured songs and visions. Both the animate and the inanimate were, in his eyes, imbued with one and the same divine life. But this aspect of his outlook is only the outer garment, so to speak, of that which is first apprehended at the superficial level. Behind the multiplicity of gods and activities, behind the mythological terminology often difficult to grasp, in the examination of which too many of the early scholars foundered, is perceptible a remarkable insight not only into the laws of the universe, both heavenly and terrestrial, but also into the depths of human nature, as well as an extraordinary vision of the oneness, the intrinsic interlinkedness of all, a vision the mightiness of which escaped the notice of nineteenth and early twentieth century Western exegesis. Such expressions as 'That One' (*tadekam*) or the 'Unborn' (*aja*) of the origins; the *ṛta* as the law, foundation and mover of the continuous development; the 'one godhood' (*asuratvam ekam*) of the devas who are the furtherers of this *ṛta*, who 'one-intentioned move together to the one purposeful accomplishment'; and the 'all one' (*viśvam ekam*) of this incessant activity which A. T. de Nicolas aptly translates as 'integral multiplicity'[3] – all these underlie, emphasise and glorify oneness:

> That which is one has developed into the all.[4]

> One whole governs the moving and the stable, that which walks and flies, this variegated creation.[5]

All issue from the One and develop in accordance with the one law which itself is rooted in the One. This is the guarantee of the one origin and the one end.

Vedic man conceived life as a web of divine forces, each object focusing, in varying degrees, the same divine force as he himself, all governed by the law of equilibrium; hence, that every deed of his had repercussions extending far beyond his present awareness. At the cosmic level *ṛta* is the law of harmony; at the human level, the law of truth, righteousness, justice; at the personal level, integrity, the manifestation of the human conscience, the silent voice that guides conduct and points to the right.

The sections on the godly and the human level, and that on the ritual, were intended to demonstrate these vistas of vast dimension that yet stand as one whole. What is merely hinted at as a passing allusion in the Ṛgveda, or expressed as a question, or as a riddle or a myth, if pieced together, is discovered to form the material for at least the broad outlines of a system of thought which found its development in the Brāhmaṇas, the Upaniṣads, the Epics and the Purāṇas.

Some scholars, like R. N. Dandekar, contest the common notion that the Veda is the fountain-head of all Indian philosophy. For him, 'historical Hinduism was affiliated to the Veda only in a formal and fortuitous manner'.[6] This is a moot point. Both the Brāhmaṇas and the Upaniṣads are expositions of, or speculations directly based upon, a number of Ṛgvedic verses – hence Ṛgvedic ideas – which nowadays are quoted as though originating from the Upaniṣads and not from the Ṛgveda where they can all be traced. This shows that certain fundamental Ṛgvedic conceptions form the fountain-spring of much of the later philosophical speculation.

A simple verse addressed to Indra in the usual style of praising the gods nevertheless contains the idea of human incapacity to reach out to godly status on his own merit, or by mere deeds, or even by sacrificial ritual (Ṛgv. VIII.70.3).

This idea is taken up and developed further in the Katha Upaniṣad thus:

> Not through much learning is the ātman reached, not through the intellect or the sacred scriptures.
> It is reached by the chosen of him because they choose him.[7]

It is not the superficial question that in the Ṛgvedic verse man is measured against Indra's greatness and found wanting, but rather that man by himself, without the divine grace, can do nothing to surpass himself. Indra is but the representative of the 'highest' man can conceive. Going beyond name and form, the Katha Upaniṣad admits the same truth and relates it to man's innermost centre, to which it no longer gives the name of Indra or Agni or Varuṇa, but simply of *ātman*, the inner self, the ultimate point wherein all forms and names dissolve.

The same holds true for the famous Ṛgvedic verse which one finds quoted again and again in later sources and in which scholars following their 'naturalistic' or 'primitive' theories or prejudices refuse to read a meaning accordant with 'later' meanings ascribed to it in Indian tradition:

> Two birds, fair-winged, close companions shelter in the self-same tree. Of the two one eats the sweet fig thereof; the other, eating not, merely looks on.[8]

The archetype of the 'bird' as the 'soul', which one finds throughout the ancient Near East, is matched by the image of the 'Tree' of life, which is also found everywhere and recurs in several passages of the Ṛgveda, among which is the fair-leaved tree beneath which revels Yama with the gods (X.135.1), and that tree which Varuṇa sustains (I.24.7) whose root is high in heaven, whose branches stream downwards, penetrating into the hearts of men. The same archetypal image occurs in the Bhagavad Gita (15.1) and in the Katha Upaniṣad (2.3.1 – 6.1) as the eternal fig-tree whose root is above, is Brahman the deathless, in whom are contained all the worlds and beyond whom none can go.[9] The multi-levelled meaning of the tree becomes a 'tradition'. The idea of the 'two' birds, also recurring in the Katha (1.3.1) and the Muṇḍaka Upaniṣads (3.1.1), is explained as the perishable and the imperishable, the light and shade, the personal and the impersonal, the mortal and immortal, the actor and the observer, the pygmy self of man and his divine Self. Why can not the original meaning of these archetypal images be traceable to the Ṛgveda, the sacred book of the Aryans, the fountain-source of Hindu tradition?

A magnificent image of the all-pervasive spirit that manifests variously as the swan (*haṃsa*) of Ṛgv. IV.40.5, that dwells in the light, in the inner space, in the home, on the altar, everywhere, finds its replica in the Katha Upaniṣad (2.2.2. – 5.2) as elsewhere and heralds the Upaniṣad doctrine of the underlying oneness of all things. But this underlying oneness is already sung in no uncertain terms in the famous Ṛgvedic hymn to *vāc*:

> Through me, whoever eats food and whoever sees,
> whoever breathes and whoever hears what is said,
> He dwells in me, though he knows it not (X.125.4).

The image of the thousand-headed, thousand-armed, thousand-eyed Puruṣa of the famous *puruṣa* hymn (Ṛgv. X.90.1–4), the all pervading spiritual essence, is repeated again and again in the later scriptures, and the Śvetāśvatara Upaniṣad, supposedly a much later work and one that stands by itself in its monotheistic tendency, uses both the image of the thousand-headed *puruṣa* of the Ṛgveda and that of Brahmaṇaspati who in the same Veda is represented as whirling his arms as a smith in the action of forging the worlds, in order to describe Rudra, its 'one God'.[10] The underlying idea remains the same, names and forms vary.

Finally, prayers for freedom from sin, found throughout the Ṛgveda and the Atharvaveda, find their echo in the Īśā Upaniṣad:

> O Agni lead us through the path of righteousness to life's fullness,
> O God omniscient

Preserve us from deceitful evil!
On thee we lavish our highest praise (18).[11]

It may be thought that these are a handful of examples, not enough to make a case. But the point at issue does not concern a plurality of examples, but the underlying 'ideas', the archetypal images and trends of thought that continue, develop and change only outwardly. These form the foundation of philosophical themes typical of the later Indian speculations.

Three great conceptions underlie the many-faceted presentation of Ṛgvedic religious life: the one law, the one truth which in the human context is integrity-integration, and the one sacrifice. These are the three fundamental aspects of the vision of *ṛta*. We have called *ṛta* the cosmic order, the universal harmony because on viewing these three main facets of *ṛta*, the keynote has ever been harmony, harmony as the expression in the manifest of the unmanifest, the Transcendent. In this respect R. Panikkar's profound statement is revealing of the inner significance of *ṛta*:

> Immanence and transcendence are intrinsically correlated and are possible only when held together in mutual tension.[12]

This holding together in mutual tension is the essence of *ṛta*: the correlation of immanence and transcendence spells the correlation of all opposites and their being held in mutual tension, in equilibrium. This is the cosmic order.

The Ṛgvedic vision of *ṛta* is one of cosmic integrity: all the various kingdoms of nature, visible and invisible, and their denizens work together in concerted action, whether consciously or unconsciously, towards a common purpose, towards establishing in all the spheres of manifested existence a perfect wholeness, ordered activity, oneness, the reflection of the Transcendent. As we have seen, underlying the multiplicity of the gods is the oneness of their godhood and their action. Underlying the many appearances and disappearances of things and creatures, from the humblest mineral and blade of grass to the proudest human and the most glorious god, is the one law of transformation which activates, organises, changes and keeps in balance the opposing forces and various tensions that work in and contribute to the unfoldment of the universe. Everything is in constant motion, on the wing as it were, but this very constancy of movement is itself the stability, the fixity of *ṛta*. Motion and stability are united in *ṛta*. Underlying the apparent disorder of certain phenomena, the one law maintains an order which, at the higher level (the godly), is perfect harmony, which at the lower (the terrestrial) is a conflict whose resolution demands the co-operation of those intelligent units of the One intelligence, gods and men: a co-operation which is movement in togetherness, a working with law and therefore on the side of power, the two forces of conflict or resistance and of co-operation or unfoldment finding their ultimate resolution at the level of the gods, where harmony is achieved and concerted action a fact, hence oneness of purpose, the pattern in the heavens. As the lower is but a blurred reflection of the higher order, the dynamics of *ṛta*, in our space-time dimension, nevertheless reveal a cosmos (in the heavens as in nature) which is the phenomenal expression of the transcendent *ṛta*. The cosmic order is constantly being mirrored on earth in spite of all.

At the human level the establishment of harmonious relationships in society is man's answer to the cosmic order: it is the expression of his understanding of it. Such relationships imply a common code of ethics, justice and honour, acceptable to all. Moral codes and sacrificial rites are reflections of man's response to the law inherent in his deepest nature, his conduct and his participation and commitment being indicative of his understanding of the cosmic order, modes of enabling him to realise the inner significance of cosmic life and his role therein. Man first submits to the Law as reflected in the social and moral order established by the wise ones who have received the vision and understood its meaning, and then re-enacts it in his ritual worship. The power of the latter lies in its calling men together for a holy purpose and in this togetherness making it easier for men to enter into a higher state of awareness, of communion with the loftier powers, a state of illumination. The drawing more closely together of men and gods is thereby furthered and the law of harmony deliberately actuated by man in this sacred work. In conforming to this universal harmony man finds his true self, meaning and purpose. His achievement of integrity heralds his integration in the cosmic order. In this attainment lies hidden the elixir of life, *amṛta*, which is the very core of *ṛta*, immortality, the essence of that which underlies the harmonious working of all things: hence *ṛta* viewed as the inherent rhythm in the law of transformation whereby all is organised into a harmonious whole, the supreme order, law, truth. All this is hinted at in many verses scattered about in the Ṛgveda and quoted throughout the present study. These, when pieced together, build up a vision of the universe that reveals such insight as has been here summarised.

The dynamic order of reality as the eternal transformation of the divine life, exemplified not only in the *puruṣa sūkta*, but also in the Agni hymns, implies the entering of this divine life into more and more restricting conditions of existence where it is buried – the stranglehold, so to speak, that material forms have over the spiritual essence that seems to die or disappear within them. This is the perpetual sacrifice whereby the universe is kept in existence, on the move, as it were. There can be no manifestation without change, without movement, the ceaseless swinging between separation and union. The love and strife of Empedocles find here their ancient Indian observation and formulation. This is the law of becoming, of life, *ṛta*, and hidden within that law is the eternal sacrifice, the immanent and transcendent *puruṣa* immolated. The fire, life-sap of all things, hidden in the waters, in the rock, in the trees, in humans, can only be brought to birth through friction, which in human terms is conflict. The law of transformation is thus revealed as equivalent to the perpetual sacrificial self-offering of the *puruṣa*.[13] This equivalence, which is not explicit but fully implicit in the Ṛgveda, is one of the most arresting conceptions in the Vedic vision of *ṛta* as sacrifice, *yajña*, one of its far-reaching conclusions. A human life that is a perpetual self-offering to the divine life that courses within itself, as within all things, is a harmonised life, for it has integrated itself in the vaster life of the universe which itself is an eternal self-offering; where there is no integration, separation, enmity, strife and unhappiness prevail. Harmony in a human being is a reflection of the cosmic order. All this is summed up, in typical Ṛgvedic fashion, in the cryptic formula addressed to the All-Creator:

> Sacrifice thyself for thine own exaltation.[14]

From such beginnings the doctrine of sacrifice found its full development in the period of the Brāhmaṇas.

The vision of *ṛta*, which in the light of the above interpretation may be called the vision of cosmic wholeness, was to split up into three main avenues of speculative thought: *dharma-karma*, *kalpa-yuga* and *brahman-ātman*. From this standpoint we can maintain that that which was peculiarly Ṛgvedic, namely the vision of cosmic order, of wholeness, of harmony, of integration, becomes the fountain-head of the subsequent, peculiarly Hindu doctrines of *dharma* and *karma*, of the law of cycles and of the essential oneness of all creatures with their Ultimate Source.[15]

The roots of the doctrine of *karma* and *dharma* are embodied in the two aspects of *ṛta*: *ṛta* as (1) the sacrifice representing both the work (*apas* or *karman*) of the universe's intelligent units, gods and humans; and as (2) what 'supports' that same universe, the basic law according to which all things move. Furthermore, the action of receiving and returning inherent in the sacrifice proclaims the law of equilibrium which sustains the universe. The doctrine of order based on equilibrium thus developed, in the moral sphere, in the doctrine of *karma*, the interconnectedness of all actions and their counterbalancing of one another – as ye have sown so are ye reaping, as ye are sowing so will ye reap – and that of *dharma*, of one's duty-vocation-function-stand in life, the regulations of conduct, codes of ethics, each appropriate to, or in strict accordance with, one's inherent truth, the innermost law of one's being, *svadharma*[16] – a doctrine which was to be further amplified in the theory of the *varṇas* and *āśramas*. *Ṛta* in its cosmic sense becomes specialised as *dharma* in a social or human sense, but a *dharma* not imposed by anyone from outside, a *dharma* seen as the inner law which governs all and which, if carried out by man in his life, results in harmony, in establishing on earth the divine harmony. In the words of the Mahānārāyaṇa Upaniṣad, *dharma* 'embraces the whole world'.[17] It is the world's foundation. Nevertheless it may be said that the great vision of cosmos as peculiar to the ṛṣis of the Ṛgveda becomes more and more dimmed and compartmentalised until the whole focus of attention is concentrated on the individual and society. Man emerges as his own most proper study and the vision of cosmic integrity disappears into the vision of Man.

It was inevitable that the conception of *ṛta* as that law which underlies the cyclic recurrence of all natural phenomena should develop further into the doctrine of the cyclic manifestation of the universe and of man, of projection and absorption, of birth, death and rebirth: in the macrocosm, manifestation and dissolution (*sṛṣṭi* and *pralaya*), the great *kalpas* and *mahā yugas* as propounded in the *Purāṇas* and the *Mahābhārata*,[18] in the microcosm, *punarjanman*, rebirth, first clearly explained in the Upaniṣads. Of these two doctrines only a few references can be found in the Ṛgveda,[19] but they are sufficient pointers both to the future of the teachings and to the fact that such ideas were already prevalent in Ṛgvedic times. Thus 'the plants that were born of old three ages before the gods' and the prayer of obeisance to the gods, the 'mighty and the lesser', the 'elder and the younger', imply a progressive formation of the earth according to cycles and a progressive development or coming into action of its denizens. Thus also the 'successive existences' (*anūcīnā jīvitā*) which Savitṛ grants man (IV.54.2), which, contrary to prevalent scholarly opinion,[20] shows far more than a 'trace' of the doctrine of rebirth.[21] That this doctrine of *karma* and *punarjanman* was regarded as a secret one may be gathered from the Brhadāraṇyaka Upaniṣad.[22] This would account for the scarce references to it in the Ṛgveda.

The vision of *ṛta* as cosmic integrity was eclipsed during

the period of the Brāhmaṇas, but only to re-emerge via its cognate term, *satya*, which thereafter assumed precedence over *ṛta*. The Ṛgvedic distinction between *ṛta* and *satya* was lost sight of, the two ideas coalescing, *satya*[23] being kept as the term for reality, truth, and *ṛta* being dropped: so that from the notion of *ṛta* as cosmic order, the one law, basis of all structures, to *satya* the truth, the one reality, to the identification of the latter with the power inherent in the invocative word, *brahman*, the creative force, to the exaltation of that power as what lies at the root of all, hence the source of all things, the truth, the steps are foreseeable and evidence a consistent line of thought and enquiry, all heading towards finding the ultimate principle.

> *Satya*, the luminous, lofty, cherished, all-conquering truth, established on the unshakeable foundations of heaven (X.170.2ab)

becomes the one reality of the Brāhmaṇas. The knowledge of the truth of the threefold gnosis, the *ṛk*, the *yajuḥ* and the *sāman*, gave the devas the power to conquer the asuras.[24] Knowledge of the truth is power. *Satya* is *brahman* which the sacrifice expresses. If *satya* is the foundation of the universe,[25] *satya* is itself founded on *brahman* which in turn is based on *tapas*.[26] Such statement is of paramount importance for it proclaims *satya* as based upon the ultimate power, the invocative hence creative power, *brahman*, which itself can only be evoked and realised through *tapas*.

The identification of *satya* and *brahman* is found in two passages of the Śatapatha Brāhmaṇa, one implicitly via worship – *yajña* being essentially the expression of *brahman*, as indeed is stated in Śat. Br. 2.1.4.10, so by implication *brahman* = *satya*:

> Worship above all is truthfulness[27]

and one explicitly:

> Verily that (fire) is of the brahman. With the brahman it is set up. The brahman is speech . . . the brahman is the truth and the truth consists in those same (three)

> mystic utterances: hence his (fire) is established by means of the truth.[28]

Fire is that burning energy which ever spurs onward and uplifts thought and feeling, blends them into aspiration and relentlessly pursues its quest: contact with the transcendent. This is the power of *brahman*. *Brahman* is the irresistible invocation: it takes form as the word that bears fruit; it becomes the word made flesh. That which is achieved thereby is the expression of that which is, hence *satya*: *brahman* is *satya*.

This identification was to be taken up and amplified in the Upaniṣads:

> *Satya* is the name of *brahman*[29]
>
> *Brahman* is *satya*[30]
>
> The wise invoke him as *ṛta*, as *satya* as highest *brahman*.[31]

Cosmic order, truth, ultimate principle, these are three stages in Vedic man's search for truth, three stages during which he searched along slightly different paths and used different terminology, but which all had the same end in view.

The Taittirīya Upaniṣad's summary could be considered as amplification of and commentary on Ṛgveda 1.67.2:

> He who knows *brahman* as the real (*satya*) as knowledge (*jñana*) as the infinite (*ananta*) laid in the secret place (of the heart) in the highest heaven, that one obtains all desires together with the all-knowing Brahman.[32]

The religious experience of the ancient seers of India, such as it may be reconstructed from their visionary hymns, thus became the basis of doctrines whose origin can be traced to the hints scattered in the Ṛgveda and whose testimony is the fountain-spring of later religious speculation. The fundamental idea of the one law that constitutes the foundation of life, the very structure of the universe and pulsation of the One Absolute Principle from whom proceed the many, first as a duality, then as a plurality, in ordered sequence; and the possibility for man of realising, through supra-normal insight or vision and through righteous living, through the integration of his whole being, and through the ritualistic ordering in symbolic gestures, certain truths that lie beyond the scope of ordinary, mundane experience and perception; all this forms the background of the hymns of the Ṛgveda and was to be taken up, developed and expounded in a different language in the Brāhmaṇas, the Upaniṣads, the Vedanta Sūtras, the Epics and the Purāṇas.

Notes

Abbreviated bibliographical references appear in full in the bibliography.

Introduction

1 J. Gonda, *The Vision of the Vedic poets*, 1963 (henceforth to be referred to as *Vision*).
2 A. C. Bouquet, *Hinduism*, 1966, p. 28.
3 R. Panikkar, *The Vedic experience*, 1977, p. 162.
4 Cf. Gonda, *Vision*: 'Man has, the Indians always believed, to acquire visionary knowledge of being, of the truth . . . of eternal values and not merely think with his brain' (p. 19).
5 For this reason all quotations from the Ṛgveda in the main body of the chapters will be noted down without the word Ṛgv. before, e.g. X.11.5c meaning book X, hymn 11, verse 5, third part. All other quotations will have the specific name of the scripture to which they belong, e.g. Athv. V.12.6.
6 Veda in the singular means divine science or gnosis; Vedas in the plural refers to the four collections of hymns, and often also the Brāhmaṇas and Upaniṣads.

Chapter 1 The means of knowledge

1 Cf. W. N. Brown, 'Theories of creation in the Rig Veda': 'In the hymn RV.X.129 . . . both the will of a deity and the power of the sacrifice are ignored, tacitly rejected, and instead there appears the idea of a single principle from which our entire universe is evolved.' This shows complete ignorance of the significance of *tapas*; will, creative urge, as well as sacrificial offering are inherent to the meaning of *tapas*.

2 The question of the world emitted by sound is a study in itself into which we cannot enter, beyond referring to the sections on Brahmaṇaspati and Soma.
3 Cf. Ṛgv. X.31.7 and X.81.4.
4 *Vision*, pp. 68–9. Concerning *dhītiḥ* Gonda writes: the 'suffix (ti) helps to form words expressing that the idea conveyed by the root manifests itself and is realised as an actuality' (p. 171).
5 H. L. Ellison, *Men spake from God*, London, 1952, p. 14. Cf. 'The prophet is God's spokesman. Speaking for God may involve foretelling the future . . . but this is secondary, not primary' (p. 14).
6 Ibid., p. 14.
7 Ṛgv. X.124.9 and cf. X.27.19. Cf. the vision of the four creatures and the wheels in Ezekiel 1:5–28; 10:10–14.
8 *Essai sur le mythe des Ṛibhavas*, Paris, Benjamin Duprat, 1847, p. 39.
9 Ṛgv. X.129.4.
10 Ṛgv. I.24.6. Cf. Psalm 139: 7–10.
11 Cf. Griffith's translation of Ṛgv. I.146.4, where the word *hṛd* 'heart' is not translated.
12 *Vision*, p. 281.
13 Brahman in the Ṛgveda means prayer, mantra, the divine *word*, the power of evocation. See J. Gonda *Notes on brahman*, Utrecht, 1950; L. Renou, 'Sur la notion de brahman' in *Journal asiatique*, tome 237, 1949, pp. 7–46; J. Miller, *The Vedas*, 1974, pp. 46–50. Also R. Panikkar, *The Vedic experience*, 1977, pp. 103 ff. and 782.
14 Ṛgv. VII.34.9.
15 Ṛgv. III.26.8.
16 Ṛgv. I.67.2.
17 Ṛgv. IV.11.2. For an explanation of this verse, see Gonda, *Vision*, p. 51.
18 *Vision*, p. 272.
19 Yjv. 34.3 and 4. Cf. Jñaneśvarī 8.87, 'Light is one of the chief forms in which God reveals himself.'
20 Ṛgv. IX.97.34b.
21 Ṛgv. IX.76.4ab.
22 Ṛgv. VII.76.4cd.
23 Ṛgv. I.164.37 Griffith's translation.
24 Ṛgv. X.5.1.
25 Ṛgv. I.164.21cd.
26 This question is dealt with in more detail in *The Vedas*, J. Miller, 1974, p. 50 ff. We are here only giving a summary of the whole problem.
27 Ṛgv. V.81.1b: *yuñjate mana uta yuñjate dhiyo viprā*. Cf. Ṛgv. VIII.13.26b: *dhiyam manoyujam*.
28 Ṛgv. VIII.42.3ab. The words *kratu* and *dakṣa* have been translated as 'resourcefulness and dexterity' in accordance with J. Gonda's remarks in *Vision*, p. 111.
29 Ṛgv. X.67.1ab. Cf. Ṛgv. VIII.51.4.

30 Ṛgv. VIII.12.10 and VII.34.8b. Cf. Ṛgv. V.2.11; I.61.4,5; 171.2.
31 Ṛgv. III.39.2. The apparent contradictions are explained by J. Gonda in *Vision*, p. 77.
32 Ṛgv. I.105.15abc. Trans. in accordance with J. Gonda's remarks in *Vision*, pp. 277–8.
33 Ṛgv. I.130.6b and III.38.1.
34 Ṛgv. VIII.59.6. Trans. in accordance with J. Gonda's remarks in *Vision*, p. 211. Cf. X.154.4.
35 Ṛgv. X.67.2.
36 Ṛgv. IV.1.14d.
37 Ṛgv. I.71.2c.
38 Ṛgv. VIII.3.18ab.
39 Ṛgv. X.154.4abc.
40 Ṛgv. I.91.1cd.
41 Ṛgv. I.109.3ab. Cf. also VIII.6.10.
42 Ṛgv. I.87.5a.
43 Ṛgv. VIII.6.10.

Chapter 2 The meaning of the word *ṛta*

1 H. Lefever remarks that the '. . . root √ṛ has two primary groups of meanings. It signifies "to move" and (through movement) "to fit or arrange". Grassman derives *ṛta* from the second group and defines it as primarily "what is firmly fixed" – "the unalterable Order" – coming later to mean "what is fitting, proper, right" ' (*The Vedic idea of sin*, 1935, p. 1).
2 'The Zodiac – the physical basis of Ṛta (Cosmic Order)', V. M. Apte, *Deccan College Research Institute Bulletin*, vol. 5, 1943–4, p. 183.
3 Lefever, op. cit., p. 1.
4 Thus J. M. Koller stresses this static aspect: 'The concept of ṛta is most often used in the Ṛgveda to signify the unchanging order of the highest reality which is the source of all order in the universe' ('Dharma: an expression of universal order', *Philosophy East and West*, vol. 22, Apr. 1972, p. 135). Cf. also A. A. Macdonell, *Vedic mythology*, 1897: 'the Cosmic order or law prevailing in nature is recognised under the name of ṛta. . . . The same word also designates "order" in the moral world as truth and right and in the religious world as sacrifice or "rite" ' (p. 11). Cf. Dumézil, 'Ordre, fantaisie', 1935, p. 140. Cf. J. Gonda, 'Some notes on the study of ancient Indian religious terminology', HR, vol. 1, no. 2, 1962, p. 257.
5 'Un hymne à énigmes du Ṛgveda.' In *Journal de psychologie normale et pathologique*, vol. 42, 1949, p. 266.
6 Lefever, op. cit., p. 5.
7 *Lectures on the origin and growth of religion*, London, 1878, p. 245.
8 In 'The Ṛgvedic antecedents of the Dharma-pāśa of Varuṇa in the

Mahābhārata', *Deccan College Research Institute Bulletin*, vol. 5, 1943–4, pp. 167–96.

9 Ibid., p. 180.

10 Ṛgv. I.124.3c = V.80.4. Max Müller takes a completely different viewpoint: 'The path of Rita occurs again and again, as followed by the dawn, or the sun, or day and night, and the only way in which we can generally translate it is the path of right, or the right path' (*Lectures on the origin and growth of religion*, p. 242).

11 Ṛgv. I.123.9c.

12 Apte, op. cit., p. 185. Cf. J. Gonda who translates and comments upon the verse thus: '. . . the young woman strictly observes that particular "modality" of universal order and truth which is "located" in, i.e. which consists in, chronological order, in casu, the regular succession of days, coming, everyday, to the place of rendez-vous (*niskṛta*)' (*Dhāman*, 1967, p. 30).

13 Ṛgv. IV.5.9c; X.123.2.

14 Ṛgv. VII.60.4c.

15 This is usually interpreted as the three seasons. The seasons vary in number, either three, as the hot, rainy and cold, or five (Śat. Br. 1.2.5.10) or even six (Ait. Br. III.1.6) – *vasanta, grīṣma, vatṣa, śarad, hemanta, śiśira.*

16 In connection with this verse Keith simply mentions that 'The year is the wheel of Ṛta with twelve spokes' in an enumeration of all the facets of *ṛta* and gives no explanation (*Religion and philosophy of the Veda*, 1925, p. 83).

Chapter 3 The metaphysical origin of *ṛta*

1 Ṛgv. X.121.2c: *yasya chāyā amṛtaṃ yasya mṛtyuḥ.*

2 Ṛgv. X.129. A full exegesis of this hymn may be found in G. Feuerstein and J. Miller, *A reappraisal of yoga*, London, 1971.

3 Ṛgv. II.1. Cf. X.121.7; 88.1; I.36.5; 144.1; 143.3. In Ṛgv. II.35.2cd Apām Napāt, the alter ego of Agni is said to have generated all things.

4 Athv. XI.8.6. The word *karman* translated as 'action' is really 'what is to be done' or 'what is ordained'.

5 Ṛgv. I.25.11. Cf. also Ṛgv. VIII.41.6a and 8.

6 *Hymns from the Vedas*, A. C. Bose, 1966, p. 7.

7 'Dharma: an expression of universal order,' *Philosophy East and West*, 1972, no. 2, p. 36.

8 Lefever, *The Vedic Idea of Sin*, 1935, p. 32. A. C. Bose sums up the argument thus: '*Satya* (truth) has two aspects; first, it is integrity, and as such part of Ṛta; secondly, it is Sat or ultimate Reality to be realised through the inner vision' (*Hymns from the Vedas*, 1966, p. 8).

For Lüders 'Ṛta denotes exclusively the truth of the spoken word or thought' (*Varuṇa*, Gottingen, 1951–9, p. 635). For him *satya*, by which *ṛta* was eventually replaced, is synonymous with *ṛta*. We do not agree with this identification, at least in so far as the Ṛgveda is concerned. The verse quoted, X.190.1, shows that *satya* and *ṛta* were distinguished in the mind of the ṛṣis. As stated by J. Gonda *ṛta* and *satya* are complementary ('Some notes on the study of ancient Indian religious terminology' in *History of religions*, vol. 1, no. 2, 1962, p. 254 ff.).

9 The meaning of *asat* and *sat* translated by nineteenth century scholars as 'non-entity' and 'entity' or 'non-being' and 'being' has aroused a great deal of controversy which is far from being over. For an elucidation of the various meanings see H. W. Wallis's *The Cosmology of the Rigveda*, 1887, pp. 61–2. We do not agree with W. N. Brown's exegesis of *asat* ('Ṛgvedic equivalent for hell' in JAOS 61, 1941, p. 76, 'Creation myth in the Ṛgveda' 62, 1942, p. 85) as he practically reduces *asat* to hell. *Asat* is the not yet determined and in some instances seems to represent that which lacks the form that gives life its intelligibility for man and thus is undesirable.

10 Cf. Ṛgv. VI.47.3.4; III.38.3.

11 Thus W. N. Brown sharply defines the opposing tensions: 'There was the created, ordered, lighted world; there was the uncreated, unordered, unillumined place of dissolution beneath it. The one was ruled by the gods; he (the Vedic sage) loved it. The other was the place of destruction and the demons; he dreaded it. Aditi, personification of Benevolent Boundlessness, symbolised one; Nirṛti, personification of Malevolence and Destruction, symbolised the other. The one he called Sat, the existent; the other he called Asat, the Non-existent' (JAOS vol. 61, June 1941, p. 79). Cf. also JAOS, vol. 62, 1942, p. 89.

Asat in the philosophical hymns has nothing malevolent. It is also noteworthy that 'mighty devas emerge from the *asat*' (Athv. X.7.25). The imprecations lavished against demons in Ṛgv. VII.104.8 by Vasiṣṭha do not necessarily mean that he is sending them to hell, but rather to non-existence.

12 *Religions of ancient India*, 1953, p. 20. Renou goes on to demonstrate that terms like *manyu*, *māyā*, *yakṣa* 'have two sets of meaning, according to whether they are used of good or evil beings'.

13 Ṛgv. VI.24.5b.

14 Ṛgv. X.90.

15 Cf. Śat. Brāhmaṇa: 'Of old everything here was within Vṛtra, to wit the Ṛk, the Yajus and the Sāman' (5.5.5.1–5), these three being considered the essence of all things.

16 Cf. Ṛgv. I.36.8ab. See also *The Vedas*, J. Miller, 1974.

17 Cf. R. Panikkar, *The Vedic experience*, 1977: 'Puruṣa is not only the cosmic man; it is also the personal aspect of the whole of reality' (p. 73).

18 Cf. L. Renou: 'The idea of evil is never clearly personified as a major demon, but is represented under the multiple forms of "hostility", "violence", "resistance"' (*Religions of ancient India*, 1953, p. 19). Cf. also R. Panikkar: 'To ask ultimate questions about the why of evil implies

two very grave assumptions a) that there is something or somebody responsible for it, and b) that evil belongs to the realms of intelligibility. The Vedas do not make either of these assumptions . . . evil is taken to originate in a malfunctioning of the given structures of reality' (*The Vedic experience*, p. 461, 462).

19 'Universe in Vedic thought', in *India Maior*, 1972, p. 93.

20 *The Cosmology of the Ṛigveda*, 1887, p. 96.

21 Athv. XI.8.7. Cf. Śat. Br.: 'There are here three earths (*pṛthivī*) viz, this one, and two beyond it' (5.1.5.21).

22 P. S. Deshmukh, *Religion in Vedic literature*, 1933, p. 318.

23 'Theories of creation in the Rig Veda', JAOS, vol. 85, 1965, p. 25.

24 Ibid., p. 27.

Chapter 4 The Vedic pantheon

1 For this question of the blend of the impersonal and personal in Vedic deities see J. Gonda, *Gods and powers*, 1957.

2 Ṛgv. X.7.3. Griffith's translation.

3 *mahad devānām asuratvam ekam* (Ṛgv. III.55.1d). Cf. II.38.

4 Ṛgv. I.139.11 and X.158.1.

5 Śat. Br. 2.5.8.2.: 'He (Prajāpati) heated these three worlds, and from them, thus heated, three lights (*jyotis*) were produced – Agni (the fire), he who blows here (Vāyu), and Sūrya (the sun).' Cf. Ait. Br. 5.32 and Ṛgv. X.158.1.

6 *Les Dieux des Indo-Européens*, 1952. Also 'L'idéologie tripartie des Indo-Européens' (Collection Latomus, vol. 31, 1958, pp. 5–122) and J. Brough's answer in his 'The tripartite ideology of the Indo-Europeans: an experiment in method', *School of Oriental and African Studies Bulletin*, University of London, vol. 22, part 1, 1959, pp. 69–85. Cf. J. Gonda, *Triads*, 1976, which discusses the whole implication of Dumézil's argument. Also the same author's 'Some observations on Dumézil's views of Indo-European mythology' (Mnemosyne VI.13, Leiden, 1960, p. 1 ff.) and 'Dumézil's tripartite ideology. Some critical observations', *Journal of Asian Studies*, 1974, p. 139.

7 As noted by A. A. Macdonell, the word *dakṣa* occurs more often as an adjective meaning 'dexterous, strong, clever, intelligent' from which he concludes that 'the name of the personification therefore appears to mean the "dexterous" or "clever" god' (*Vedic mythology*, 1897, p. 46). L. Renou qualified *dakṣa* as a 'complementary term of *kratu* indicating practical efficiency, the concrete realisation conferred by the possession of *kratu*' (EVP IV, 1955, p. 31).

8 A. B. Keith dismisses Dakṣa as: 'the word means no more than "clever" and there can be no doubt that he is the product of priestly ingenuity' (*Religion and philosophy of the Veda*, 1925, p. 100). Cf. P. S. Deshmukh, *Religion in Vedic literature*, 1933, p. 232. Similarly, for H. W. Wallis

Dakṣa is 'an unimportant deity whose only characteristic in the Ṛgveda is his fatherhood of the gods' (*Cosmology of the Rigveda*, 1887, p. 45), who derived his personality through the same process as that which created the goddess Śavasī out of such epithets as 'son of strength' ascribed to Indra. It is no criterion to judge the importance of deities on the number of times they are invoked. The deities mentioned most are those that stand in closer contact with mankind, e.g. Agni, Indra, Soma, Savitṛ, the Aśvins, Varuṇa and Vāyu. But the function in creation of those who are more remote may be out of all relation to their specific assistance to man.

9 Ṛgv. VIII.25.5ab; III.62.17; I.23.4; V.10.2; VII.66.2.

10 For J. Gonda Dakṣa is an example of those cases 'in which a definite "power" is conceived as an "impersonal potency" and sometimes as a "divine person" . . . dakṣa . . . is personified as "the Clever or Dexterous One" i.e. "cleverness" or "dexterity" ' (*Gods and Powers*, 1957, p. 5). Cf. J. Muir's doubt as to 'the propriety of taking Dakṣa (in Ṛgv. VI.50.2 and VII.66.2) . . . to represent a person, from the fact that in Ṛgv. VIII.25.5 Mitra Varuṇa are not only called the "strong sons of Dakṣa" . . . but also the "grandsons of mighty strength" ' (*Original Sanskrit Texts*, 1870, vol. 5, p. 52). The fact is that the personification of Dakṣa is only incipient in the Ṛgveda and is developed and completed in the Purāṇas.

11 Ṛgv. I.139.2ab.

12 For the etymology of the word see A. A. Macdonell (*Vedic mythology*, 1897, p. 121; *Vedic grammar*, 1910, p. 122); V. M. Apte ('Is Diti in the Ṛgveda a mere reflex of Aditi' in *Bharatiya Vidya* 9,1949 p. 15–16). The word, like *dakṣa*, is also used as an adjective.

13 *aditer-upasthe* (Ṛgv. X.5.7). Cf. Athv. VII.6.1–4.

14 Ṛgv. I.89.10. For an assessment of Aditi see K. Werner 'Symbolism in the Vedas and its conceptualisation' in *Numen*, vol. 24, fasc. 3, 1977.

15 *anarva* (Ṛgv. II.40.6c); *uruvyacāḥ* (Ṛgv. V.46.6d); *dhārayat-kṣitim* (Ṛgv. I.136.3b).

16 Ṛgv. I.24.7.

17 Ṛgv. VIII.101.16. Griffith's translation.

18 Ṛgv. I.164.42. W. O'Flaherty's translation.

19 *saya ākāśo vāg eva sā* (Jaiminiya Upa. Br. I.23).

20 Taitt. Br. 2.8.8.5.

21 Cf. J. Muir, op. cit., p. 369. Cf. P. S. Deshmukh, A. A. Macdonell, A. B. Keith, etc., who give no explanation for this phenomenon.

22 Ṛgv. III.2.3.; V.3.4.; VI.7.4.; X.88.8,9,10.

23 J. Gonda considers Virāj 'a creative principle representing also the idea of ruling far and wide, being the sum of all existence, the hypostatization of the conception of the universe as a whole' (*Four studies in the language of the Veda*, 1959, p. 155). Cf. also the same author's *Dhāman*, 1967, p. 26. Cf. Bhagavad Gita 15:16 where the two

puruṣas are referred to: the *kṣarạ* or perishable and the *akṣara* or imperishable.

24 'The sources and nature of *puruṣa* in the *Puruṣa sūkta* (Ṛgveda X.90)' in JAOS, vol. 51, 1931, p. 108–18. Cf. also his definition of puruṣa on p. 114.
   The word *puruṣa* seems, even in Ṛgvedic times, to have borne several shades of meaning, e.g. Agni, called 'puruṣa of plants' (Ṛgv. X.51.8) appears to be the 'vital part' or soul of plants, just as he is the 'immortal guest in mortal bodies'. This could point to a meaning of puruṣa which is not generally considered, viz. the soul or divine essence as the life energy that courses through all things.

25 Whether the notion of the *sat* re-merging into the *asat* was present in the *ṛṣis'* mind to appear later as the cycles of *sṛṣti* and *pralaya*, cannot be surmised. No hint of it has been found except in Ṛgv. X.190.3 where the words *yathā pūrvam* (as before) can be interpreted as referring to a previous age and thus universe. Cf. also Ṛgv. X.72.9 and Athv. XI.8.7.

26 Cf. Ṛgv. X.114.1; VII.5.4; I.154.4; I.102.8ab; 105.5ab. Cf. Athv. XII.3.20; XVI.50.2.

27 Cf. Ṛgv. II.27.8,9.

28 Cf. Ṛgv. VIII.69.3cd; I.105.5ab; V.69.1ab; VI.47.3cd.

29 Dandekar explains *antarikṣa* thus: 'From *antari* and *kṣa* (from *kṣi* "to reside") . . . also analysed as *antar* and *ikṣ* (for *īkṣ*) and explained as "transparent" . . .' ('Universe in Vedic thought' in *India Major*, 1972, p. 101).

30 '*Antarikṣa, ākāśa, kha*' (*Rocznik orientalistyczny*, tome 24, part 2, 1961, pp. 53–8). This survey of scholars' contributions to the interpretation of *antarikṣa* shows that alone the notion of realm of air, aerial ocean, aerial sphere, atmosphere, space of air, has been investigated.

31 Śat. Br. 5.1.5.21. Cf. Ṛgv. VII.104.11; I.34.8; and Athv. IV.20.2: 'Three heavens and three earths'. Also Ṛgv. VIII.41.9 where Varuṇa rules over the 7 worlds.

32 *Triads*, 1976, pp. 53–4.

33 S. Kramrisch, 'The triple structure of creation in the Ṛg Veda', *History of Religions*, vol. 2, no. 1, 1962, pp. 140–1 and 149.

34 Ṛgv. III.54.5.

35 *sujyotiṣaḥ . . . dakṣapitṝn* (Ṛgv. VI.50.2a).

36 Cf. Ṛgv. VII.66.2. The Tait. S. 1.2.3.1. calls them 'mind-born deities, mind-yoked, highly intelligent sons of Dakṣa'.

37 Cf. A. D. Pusalker 'Purāṇic cosmogony' in *Bharatiya vidya*, vol. 2, part 2, 1941, p. 178; R. N. Dandekar 'Universe in Vedic thought' in *India Major*, 1972, p. 98.

38 Ṛgv. I.103.2. Cf. II.15.2 and 3; X.54.3cd; IV.56.3.

39 *prapūrvajate pitarā* (Ṛgv. VII.53.2).

40 Ṛgv. X.12.1.

41 Griswold, *The religion of the Rigveda*, 1923, p. 100.

42 Ṛgv. VII.53.2ab; IV.56.5; V.43.2, etc.

43 *Four studies in the language of the Veda*, 1959, p. 168.

44 *The Vedic idea of sin*, 1935, p. 12.
45 Ṛgv. V.85.5, Griffith's translation. Cf. also V.85.1.
46 Kramrisch, op. cit., p. 143.
47 VIII.41.10c: *sadhāma pūrvyam mame*; VIII.41.4c: *sa mātā pūrvyam padam*.
48 Cf. Naighantu 1.12: water, *ṛtasya yoniḥ*. As remarked by F. B. J. Kuiper, 'the question whether Varuṇa was "primarily" a god of the waters, a chthonic, or a celestial god has been the subject of fierce disputes'. In the light of his structural interpretation 'such disputes are senseless' ('The bliss of Aśa', *Indo-Iranian Journal*, vol. 8, 1964, p. 115).
49 Cf. Ṛgv. X.72.2–3 and Athv. 11.8.10.
50 Ṛgv. IV.42.4cd.
51 Ṛgv. VII.87.5ab, Griffith's translation.
52 Ṛgv. VIII.41.10c.
53 Ṛgv. IV.42.4b.
54 Ṛgv. VII.49.3.ab.
55 Ṛgv. VII.49.3c; IX.90.2; VIII.69.11,12.
56 Cf. Ṛgv. I.25.5,16; VIII.90.2; VII.34.10.
57 Ṛgv. II.12.2. Cf. Ṛgv. VI.17.7; II.13.7; 15.2,3; I.103.2; VII.98.3.
58 Ṛgv. III.49.4. Cf. Ṛgv. IV.26.1; VIII.82.4; 3.6; II.19.3; 12.7.
59 Cf. Ṛgv. VI.20.2; VII.21.7.
60 *Vedic mythology*, Strasburg, 1897, p. 63.
61 Ṛgv. X.50.4; Cf. I.80.1; VII.19.11; III.34.1.
62 Ṛgv. VI.44.23. Cf. also III.34.7,8; VII,98.3d for Indra as winner of the light.
63 Cf. Ṛgv. IX.72.7; 77.1; 47.3.
64 Ṛgv. IX.47.3.
65 See *The Vedas*, J. Miller, 1974, p. 36.
66 *apām garbha* (Ṛgv. I.70.3); III.1.12,13; 5.3; 'embryo of the two worlds' X.1.2.
67 Ṛgv. X.5.7.
68 Ṛgv. X.45.1abc. Adapted from Griffith's translation. Cf. X.46.9.
69 Cf. Ṛgv. X.2.7; III.22.2. Agni, also like Brahmaṇaspati 'with effective utterances fixed the sky' (Ṛgv. I.67.3b), and like Indra spreads the mid-region.
70 Kramrisch, op. cit., p. 155. See also G. Feuerstein and J. Miller *Reappraisal of yoga*, 1971, pp. 121–51.
71 Ṛgv. IX.5.9; I.13.10.

Chapter 5 The working of the Law

1 For the various interpretations of this word see J. Gonda, *Dhāman*, 1967, and L. Renou, *EVP*, vol. V, p. 73. Also Dumézil 'Ordre, fantaisie', 1955, p. 140.
2 Gonda, op. cit., p. 35.
3 *sa dhāma pūrvyam mame* (Ṛgv. VIII.41.10c).
4 Ṛgv. X.124.3b and 122.3a.

5 Cf. Ṛgv. X.181.2ab.

6 *dharman divo dharune satyam arpitam* (Ṛgv. X.170.10c).

7 The connection between *dharma* and *ṛta* is well explained by J. M. Koller: 'The ordering or regulation of relations between events is accomplished by *dharma* . . . *Dharma* is the expression of cosmic ṛta in human life, providing the identity of the individual reality with the higher reality . . . all ordinary human dharma is only an aspect of the universal dharma, and is justified not in itself, but only in the function of the universal dharma, the ṛta of the Ṛgveda' ('Dharma: an expression of universal order' in *Philosophy East and West*, vol. 22, 1972, p. 141).

8 For the various opinions as to the derivation of *vrata* see V. M. Apte, 'All about "vrata" in the Ṛgveda', in *Deccan College Research Institute Bulletin*, vol. 3, 1942, p. 409, also vol. 5, 1943–4, p. 179; W. Whitney, JAOS XI, 1884, pp. ccxxix–ccxxxi, 'On the etymology of the Sanskrit noun vrata'; Bergaigne *La religion védique*, tome III, 1883, p. 212; H. W. Wallis, *Cosmology of the Rigveda*, 1887, p. 99.

9 *ṛtasya devā anu vratā guḥ* (Ṛgv. I.65.3b). Cf. Ṛgv. V.67.3.

10 This poses a further question about the meaning of *antarikṣa*. Is it a kind of heaven?

11 Ṛgv. VIII.86.5ab. In this particular case the 'peak of *ṛta*' could be the circle of the zodiac. Cf. V. M. Apte 'The Zodiac – the physical basis of Ṛta (Cosmic Order)', *Deccan College Research Institute Bulletin*, vol. 5, 1943–4, p. 186–7.

12 *anu vratam catasi* (Ṛgv. III.61.1cd).

13 *janayantī daivyāni vratāni* (Ṛgv. VII.75.3d).

14 Ṛgv. VIII.41.7. Cf. V.67.3; VI.67.5.

15 *ṛtapeśas* (Ṛgv. V.66.1).

16 Ṛgv. I.25.11. All that Varuṇa represented in Vedic times is epitomised in a hymn of the Atharvaveda (Athv. IV.16) of which R. Roth remarked 'There is no hymn in the whole of Vedic literature which expresses the divine omniscience in such forceful terms as this' ('Abhandlung über den Atharvaveda' in *Tübinger Universitäts-schriften*, 1856, p. 30, as quoted in 'Asura Varuṇa' by R. N. Dandekar (*Bhandarkar Oriental Research Institute Annals*, vol. 21, April–July 1940, p. 158, note 1)).

17 No better summing up of the function of the law terms can be found than in H. Lefever's *The Vedic Idea of Sin*: '. . . the terms *vrata, dharma* and *dhāman* stand for the ordinances, formulated (generally by the Gods) in accordance with ṛta, for the application or realisation of ṛta in the world. These ordinances can be said to "belong" to the gods in a sense which is never true of ṛta. . . . The Gods "found" or "establish" the dhāman, as they establish the physical world, or as a poet constructs his song. Similar is the case with *vrata* and *dharman*. The Gods do not create ṛta, however. It would be more correct to say that they find ṛta' (p. 12).

18 Ṛgv. V.68.4 (cf. I.136.4d, VII.62.3); V.63.1ab,7c.

19 *The Vedic idea of sin*, p. 19; p. 12; p. 14.

20 In *Philosophy East and West*, vol. 16, July–Oct. 1966.

21 Ibid., p. 144.

22 J. Gonda remarks that 'Creative power was indeed attributed to those vested with asuraship already in the Ṛgveda: 2.35.2. "Apām Napāt created by the greatness of his asuraship all being" cf. also 6.30.2' (*Four studies in the language of the Veda*, 1959, p. 157). According to R. N. Dandekar, 'The asura . . . is the being who possesses the highest amount of *asu*, the occult power-substance and therefore commands the greatest occult power. Varuṇa is *asura* par excellence' ('Asura Varuṇa' in *Annals*, Bhandakar Oriental Research Institute, vol. 21, Apr.–July 1940, p. 180). Cf. also *Some aspects of the history of Hinduism*, 1967, p. 39, and A. A. Macdonell, *Vedic mythology*, 1897, p. 156–7.

23 Ṛgv. VI.20.2, Griffith's translation. See also II.23.2 for Brāhmaṇaspati; III.3.4 and VII.30.3 for Agni; X.55.4 for Uṣas; IV.53.1 for Savitṛ; I.35.7 and 10 for Dyaus etc.

24 W. N. Brown differentiates the classes of the asuras as the Adityas and the Dānavas and oversimplifies and crystallises into rigid patterns of opposition the rather fluidic Vedic conception of the universal opposing forces of expansion and constriction, in fact, of involution and evolution. 'The creation myth of the Rig Veda' in JAOS, vol. 62, 1942, p. 88. Cf. also A. de Nicolas, *Four-dimensional man*, 1971, pp. 111–14. The gods' battling with the *asuras* is mentioned twice in the last book of the Ṛgveda: X.53.4 and 157.4. Certain asuras opposed to the Adityas are named: Pipru (X.138.3), Namuci (X.131.4), Varcin (VII.99.5), Śambara and Śuṣṇa.

25 A. A. Macdonell summed up the position thus: 'To the Vedic poets asura must have meant "possessor of occult power" and as such would have been potentially applicable to hostile beings. In one hymn of the RV (X.124) both senses seem to occur. Towards the end of the Rigvedic period the application of the word to the gods began to fall into disuse' (*Vedic mythology*, 1897, p. 156–7).

26 R. N. Dandekar derives *māyā* from *mimāti* or *mimīte*, meaning to 'lay down', 'arrange' or 'organise' (*Some aspects of the history of Hinduism*, 1967, p. 39). J. Gonda argues that the identification of māyā with magic is questionable, it being more of a 'force mensuratrice' enabling its possessor to create dimensional reality (*Four studies in the language of the Veda*, 1959, p. 164), but this is exactly what the term 'magic' meant to the ancients. *Māyā* has also been derived from √*mā mināti*, to change, exchange. See P. Thieme, ZDMG 95 (1941), pp. 82–116, and G. Dumézil 'Ordre, fantaisie' etc. in *Revue des études latines*, tome 32, 1954, p. 142 ff. Cf. A. A. Macdonell's *Vedic mythology*, 1897, p. 24: 'This term signifies occult power, applicable in a good sense to gods or in a bad sense to demons. It has an almost exact parallel in the English word "craft", which in its old signification meant "occult power, magic",

then "skilfulness, art" on the one hand and "deceitful skill, wile" on the other.'

27 'The doctrine of māyā' in Agra University *Journal of Research*, vol. 1, November 1952, p. 33. Cf. V. K. Rajnade 'Asurasya māyā in Ṛgveda' (First Oriental Conference, Poona, procs. and trans., 1920, vol. 1, pp. ix–xiii).

28 Cf. J. Gonda's refutation of L. Renou's argument as to the 'gradual coalescence of two originally different words *māyā*', in '*Four studies*', 1959, p. 190.

29 Cf. Ṛgv. V.85.5; I.159.4. Cf. III.53.8 for Indra's transformations.

30 Cf. Ṛgv. III.20.3, 27.7, etc.

31 Athv. XII.1.19.20, A. C. Bose's translation (*Hymns from the Vedas*, 1966, p. 367).

32 Cf. Ṛgv. II.11.5 and 9; I.51.4; V.30.6; 40.6 and 8. The demons however may simply be Vedic man's personification of those forces of nature that to him seemed to counteract established order. Certain phenomena in nature seemed to have been considered as going against the established norms. Thus when during daytime the sun is hidden in darkness through an eclipse, this is against the *vrata*, it is *apavrata*, as stated in Ṛgv. V.40.6. The verse is very ambiguous and may refer both to the actual sun in an eclipse and to the inner sun of illumination which Atri finds in the fourth degree of prayer. Furthermore, since planetary movements can be calculated and eclipses predicted, the latter are in fact part of the cosmic order, they cannot really be against *ṛta*, though seemingly, in the eyes of Vedic man, against one sphere of *ṛta*.

33 Cf. Ṛgv. I.51.5; 11.7; 32.4; III.53.8; V.30.6; X.73.5; 147.2.

34 Ṛgv. III.56.1ab.

35 Ṛgv. I.30.3 and 5 seems to recount a war in heaven with Indra fighting all the gods, not just the particular asuras that are his enemies. But no further reference is made anywhere else to this mythological episode which could portray certain astronomical phenomena. Cf. A. K. Coomaraswamy 'Angel and Titan' (JAOS, vol. 55, 1935, p. 384, note 14).

36 Ṛgv. VI.9.5cd. Cf. X.101.1; 126.1; VIII.27.5.

37 *svam . . . cakṛṣe bhūmiṃ pratimānam ojasaḥ* (Ṛgv. I.52.12c). Cf. III.2.2, addressed to Agni.

38 Ṛgv. VI.67.5ab. Cf. X.92.4. and VII.21.7ab.

39 Cf. X.124, where the beginnings of a change can be seen.

40 Ṛgv. X.65.7ab.

41 Ṛgv. II.35.15c.

42 Ṛgv. I.147.1d.

43 Ṛgv. IV.40.5.

Chapter 6 The meaning of *anṛta* (disorder, chaos, falsehood) and the role of dissolution (*nirṛti*)

1 'Un hymne à énigmes du Ṛgveda', *Journal de psychologie*, 1949, p. 270. Cf. also R. Panikkar, *Vedic experience*, 1977, p. 61–2: 'The cosmic pillar'.
2 Ṛgv. III.39.5 and 7a.
3 Cf. Ait. Br. 4.17: 'the cows are the Adityas'. Max Müller in his *Contributions to the science of mythology* mentions three kinds of cows: 'There were thus three kinds of cows, the real cows, the cows in the dark cloud (rain – milk), and the cows stepping forth from the dark stable of the night (the rays of the morning). These three are not always easy to distinguish in the Veda, nay, while we naturally try to distinguish between them, the poets themselves seem to delight in mixing them up' (London, Longmans, 1897, vol. II, p. 761).
4 Athv. IV.10.5. Cf. Śat. Br. 5.5.5.1–5. A. B. Keith considers this verse a 'late and absurd legend' (*Religion and philosophy of the Veda*, vol. 31, 1925, p. 104).
5 See Ṛgv. II.24; X.108.
6 L. Renou, 'Védique nirṛti', *Indian linguistics*, vol. 16, 1955, p. 11. When Indra is represented as having prospered and acquired much delight through human prayer he is asked not to be a *paṇis* (Ṛgv. II.33.3), not to withhold from man the treasure that has come to light from within him and that he can bestow.
7 Ṛgv. IV.2.15cd. Cf. O.T. Isaiah: 'And I will give thee the treasures of darkness and the hidden riches of secret places' (45:3). Cf. also Exodus 17:6; Numbers 20:11; Deut. 8:15.
8 In this respect V. M. Apte makes the pertinent remark that 'the ensnaring net of ṛta probably means this complicated and involved tangle or web of ṛta and anṛta and represents the ramifications of the nooses of Varuṇa escape from which was, however, possible by strict adherence to the path of ṛta. . . . Ṛta (or its custodians Mitra and Varuṇa) was itself *the bond that bound (the snares of) anṛta*. . . . To separate the confused strand of ṛta and anṛta in this tangled skein was no easy task for god or mortal' ('The Ṛgvedic antecedents of the Dharma-pāśa of Varuṇa in the Mahābhārata', *Deccan College Research Institute Bulletin*, vol. 5, 1943–4, p. 194). Hence the poet's predicament: 'which is your ṛta, which your anṛta' (Ṛgv. I.105.5cd).
9 Or 'those who drink from the cup of *anṛta*', as the same verse could be translated (Ṛgv. V.12.4c).
10 *The Vedic experience*, 1977, p. 462.
11 Renou, op. cit., p. 14.
12 For the Ṛgvedic hell see J. Miller, *The Vedas*, 1974, pp. 161–7. See also W. N. Brown 'The Ṛgvedic equivalent for hell', JAOS, vol. 61, 1941, pp. 76–80; R. Roth 'On the morality of the Veda', trans. W. D. Whitney, JAOS, 3, 1853, pp. 329–47.

13 L. Renou explains *nirṛti* as 'a derivative in *ti* (thus bearing a priori some dynamic value which makes it appropriate for personification) from the root (*ar*) ṛ pointing to a certain fixed order, a foreseen arrangement of time, cosmos, human or ritual activity. . . . With the prefix *nis*, *nirṛti* will inversely signify disorder – the term denoting a factor of entropy' (op. cit., p. 11).
14 *mṛtyubandhu*, *mṛtyupāśa* (Athv. VIII.8.10 and 16).
15 Athv. XII.2.3.
16 Athv. VI.63.2.
17 According to L. Renou, the problem 'concerns an entity which "hidden in the bosom of the mother, has penetrated into *nirṛti* whilst having numerous progeny". This must be the breath which, having come back to its original point, abolishes itself whilst ceaselessly giving birth to other breaths. *Nirṛti* is here the apparent cessation of activity, a rupture of organic circuit (ṛta)' (op. cit., p. 14).
18 Renou, op. cit., p. 14.
19 For B. G. Tilak (*The Arctic home in the Vedas*, 1903) *nirṛti* is more than a state of dissolution, it is a locality which he identifies with the 'West'. All the passages he quotes (I.164.32 and 33, X.114.2, 161.2, 95.14, VII.58.1) show, according to him, 'that nirṛti, or the land of dissolution and death, commenced in the West, that the sun lying in darkness travelled through the distant region (*parāvat*) and eventually rose in the east from the lap of nirṛti, and that the whole of this movement was placed not in the upper heaven, but on the other side of the vault through which the sun travelled before he entered into nirṛti. In other words, the nirṛtis extended below the earth from west to east; and since the region below the three earths is expressly mentioned in the Ṛgveda, the three Nirṛtis must be understood to mean the three regions below the earth corresponding to the threefold division of the earth or of the heaven above it' (p. 262).
20 Ṛgv. I.116.11d.

Chapter 7 The concept of sin in the Vedas

1 Ṛgv. IV.54.3: *puruṣatvatā*.
2 Ṛgv. VII.86.5 and 6.
3 Ṛgv. V.85.2.
4 Ṛgv. I.105.15cd.
5 Cf. Ṛgv. X.35.8.
6 Cf. Ṛgv. I.83.5; IV.1.13ab.
7 Cf. N. T. Math. 6:2: 'forgive us our debts'.
8 *ṛṇāni cayamānāḥ*: (Ṛgv. II.27.4d). Cf. Ṛgv. IV.23.7; X.89.8.
9 H. Lefever, *The Vedic idea of sin*, 1935, pp. 24 and 9. The role of the gods, their mediating action in the ethical order of humanity is well brought out by H. Lefever: 'The Gods . . . are "charioteers of ṛta",

guarding the transcendent cosmic law by means of their statutes. These statutes have thus their origin, not so much in the pure will of the Gods, as in the transcendent ṛta. Therefore the breach of such statutes is not so much a personal offence against the Gods as a violation of the ṛta which the Gods protect.
All wrong-doing, whether against Gods or men, is a breach of this Law and herein, rather than in the character of the Gods, lies the special nature of the Ṛgvedic idea of sin' (op. cit., pp. 21 and 23).

10 *āgas, enās, agha, pāpa, śaraṇi, duskṛta, ṛṇa*. J. Gonda ascribes to *āgas* the sense of 'pollution' in accordance with the later texts (*Gods and powers* 1957, p. 81). According to H. Griswold, 'sin itself is viewed as transgression and indebtedness' (*Religion of the Ṛgveda*, 1923, p. 125). Cf. R. Panikkar, *Vedic experience*, 1977, pp. 482–3 and H. Lefever, op. cit., p. 26, for whom *enās* is literally an act of aggression, *āgas* 'sin in its deepest and most ethical sense'.

11 R. Shamasastry, *The conception of sin in the Vedas*, Festschrift Moriz Winternitz 1865–1933, 1933, p. 141; cf. R. Panikkar, op. cit., pp. 482–3; J. Mackenzie, *Hindu ethics*, 1922; M. Bloomfield, *Religion of the Veda*, 1908.

12 Lefever, op. cit., p. 18.

13 As Lefever remarks: 'In sending either prosperity or adversity the gods were acting . . . in accordance with the ṛta' (op. cit., p. 16). Adversity is a result of the infringement of the law from which one may learn many a lesson.

14 *enās, āgas, drugha, pīḍā.*

15 *duḥkha, durita, anṛta, adharma.*

16 *The Vedic experience*, 1977, pp. 482–3.

17 Ṛgv. X.10.12,6,4; X.89.9; II.29.5.

18 Ṛgv. X.71.6.

19 Ṛgv. X.117.1,2,5,6. Griffith's translation.

20 Cf. Ṛgv. V.85.7. The ideal of *ahiṃsā* does not appear in the Ṛgveda, violence being accepted as a part of life that cannot be dispensed with. Its first mention is found in the Chāndogya Up.: 'Austerity, alms-giving, uprightness, harmlessness, truthfulness – these are one's gifts for the priests' (III.17.4). The derivative word *ahiṃsāna* is used in Ṛgv. V.64.3, qualifying Mitra who does not harm his devotees, and *ahiṃsyamāna*, being unharmed, is found in Ṛgv. I.141.5. The conception had evidently not yet developed into a cardinal doctrine.

21 Ṛgv. X.85.1; 190.1; Athv. XII.1.1.

22 Athv. XIX.44.8. Cf. also Ṛgv. I.147.5.

23 Cf. *ṛtam vada*, speak the truth; *ṛtam āśuṣāṇāḥ*, they breathe forth the truth.

24 *Oriens*, vol. 13–14, 1960–61, pp. 398–410.

25 Cf. Ṛgv. I.39.10; X.97; X.9.8; Athv. V.4; 5; XIX.69.

26 *Ethics of India*, 1924, p. 25. Cf. Ṛgv. V.82.5 and 11; 28.9.

27 Athv. XII.2.40.

28 Ṛgv. VI.74.3, Griffith's translation.
29 *La religion védique*, 1878–83, tome III, p. 158.

Chapter 8 Karma inherent in the conception of *ṛta* in the moral sphere

1 ' "When the voice of a dead man goes into fire, his breath into wind, his eye into the sun, his mind into the moon, his hearing into the quarters of heaven, his body into the earth, his soul (ātman) into space (ākāsha) . . . what then becomes of this person (purusha)?" "My dear, take my hand. We two only will know of this. This is not for us to speak in public." The two went away and deliberated. What they said was karma. What they praised was karma. Verily one becomes good by good action, bad by bad action' (3.2.13).

The question of karma is bound up with the problem of reincarnation into which we will not enter besides pointing out that this idea seems to be expressed in Ṛgv. IV.54.2. See K. Werner 'The Vedic concept of human personality and its destiny', *Journal of Indian philosophy*, 5, 1978, pp. 286–7, and J. Miller, *The Vedas*, 1974, p. 184 ff.

2 *Ethics of India*, 1924, pp. 43–4.
3 Bṛh. Up. 4.4.5.
4 See Ṛgv. VI.67.5; VIII.28.9; II.27.3; VII.61.5; I.25.13; VII.34.10; VIII.41.7; VI.51.2.
5 Ṛgv. IX.97.18ab.
6 H. Lefever, *The Vedic idea of sin*, p. 26.
7 Ṛgv. IV.5.5cd.
8 Lefever, op. cit., p. 33.
9 Ṛgv. II.28.9b.
10 Ṛgv. VII.52.5cd.
11 R. Shamasastry, *The conception of sin in the Vedas*, Festschrift Moriz Winternitz, 1865–1933, p. 141.
12 Ṛgv. VII.86.5a. Cf. O.T. Exodus 20:5: 'For I the Lord thy God am a jealous God, visiting the iniquity of the fathers upon the children unto the third and fourth generation of them that hate me.'
13 Shamasastry, op. cit., pp. 141–2.
14 This idea is beautifully expressed in Kahlil Gibran's *The Prophet*: 'And when one of you falls down he falls for those behind him, a caution against the stumbling stone. Aye, and he falls for those ahead of him, who, though faster and surer of foot, yet removed not the stumbling stone' (London, Heinemann, 1955, pp. 37–8).
15 The solidarity and interdependence of life are the theme of the legend of Śunaḥśepa, well expounded by Dr Panikkar in his *Myth, faith and hermeneutics*, 1979. He shows how the solidarity of life pervades every level of existence: in pledging his son's life Hariścanda offers his own; when the son flees the father falls ill. Śunaḥśepa 'redeems Rohita who

was not ready to give up his life.' Dr Panikkar draws the remarkable conclusion: 'This mytheme tells us that we can be completely cornered and have no other recourse but to leap into a brand new sphere that transcends the spatio-temporal individual. In more popular language, the just must pay for the sinner since they are the only ones who can pay. They are called just precisely for this, that they do not mutter out of a misplaced sense of individualistic propriety, and so do not find their fate unjust (or else they would no longer be just)' ('Myth, faith and hermeneutics; cross cultural studies', New York, Paulist Press, 1979, p. 158).

16 Lefever, op. cit., pp. 33–4. Cf. p. 14: 'The conviction of a world order, to which all things and in particular all men had to submit, deprived an extreme individualism of the right to set itself against the whole. The moral imperative thus receives its sanction from the law and order of the entire universe.'

17 *iṣṭāpūrta*. A. B. Keith qualifies this idea as 'a distant precursor of the later karman' (*Religion and philosophy of the Veda*, 1925, p. 250).

18 Athv. XVIII.3.58. This is not the place to discuss the meaning of the 'splendid body' with which the departing person is to unite. The subject belongs to Vedic eschatology and has been treated in *The Vedas*, 1974, pp. 134–61.

19 Ṛgv. X.135.5–6.

20 Ṛgv. VII.61.5cd: *druhaḥ sacante anṛtā janānām*.

21 Ṛgv. IX.73.6d: *ṛtasva panthāṃ na taranti duskṛtaḥ*.

22 R. Panikkar, *The Vedic experience*, 1977, p. 540.

23 *Karma and reincarnation in Hindu religion and philosophy*, 1927, pp. 6–7.

24 Panikkar, op. cit., p. 355.

25 Athv. XII.1.1ab.

26 Ṛgv. I.90.6ab.

27 Yjv. 20.25.

28 *An outline of the religious literature of India*, 1967, p. 35.

29 *Religion and philosophy of the Veda*, 1925, p. 35.

Chapter 9 Varuṇa and the fetters of sin

1 Ṛgv. VIII.41.8,5,6,9.

2 Ṛgv. VII.60.5.

3 Here is seen the restoration of a disrupted equilibrium, the bringing of that which has broken away back into the one harmony – the root idea of karma.

4 Ṛgv. VIII.25.24; cf. VI.67.1.

5 For the root of both words, Varuṇa and Vṛtra, see A. B. Keith, *Religion and philosophy of the Veda*, 1925, p. 100; A. A. Macdonell, *Vedic mythology*, 1897, p. 159. For a contrary view see R. N. Dandekar's 'Asura Varuṇa', p. 163. See also Bergaigne, *La religion védique*, 1878–83, tome

II, p. 200; E. Benveniste and L. Renou, 'Vṛtra et Vreragna: Étude de mythologie indo-iranienne', Cahiers Soc. Asiatique, Paris, 1934, p. 101.

6 Ṛgv. IV.42.1d.

7 They are mentioned in the Athv. X.2.31,32.

8 Cf. Ṛgv. VIII.6.10; I.164.21; Yjv. 31.18.

9 Cf. Ṛgv. I.24.13 and 15. Rudra is also referred to as having a net (Ṛgv. VII.46.6b) wherein he can catch the sinner.

10 Sin is compared to a fetter which tightens around the transgressor. Words such as *raśanā* (Ṛgv. II.28.5) rope, *dāman* (Ṛgv. II.28.6 and VII.86.5) halter, *bandhana, pāśa, setṛ* are in constant use. Hence the sinner asks Varuṇa to slacken or loosen √*viśrath*, or √*ava sṛi* cast off, √*vicṛt* untie the fettering cord, to release √*muc*, him like a thief or a calf from the rope. Cf. Ṛgv. VII.86.5.

11 Cf. Ṛgv. X.85.24.

12 Ṛgv. II.28.7ab, Griffith's translation. Cf. Ṛgv. I.24.13;5,7,9; II.28.8.

13 Cf. John 3:36; Jeremiah 10:10.

14 R. Otto, *The idea of the holy*, 2nd ed., 1950, p. 19.

15 Cf. Exodus 4:24, 'the lord met him and sought to kill him'.

16 Ṛgv. I.25.1 and 2; cf. VII.89 and X.137.1 addressed to all the gods.

17 Ṛgv. I.24.11cd.

18 Ṛgv. X.83,84; Athv. IV.31;32 with variations.

19 Ṛgv. X.129.3.

20 Ṛgv. X.83.3ab and 4. Griffith's translation. For further elucidations on this aspect of fire see D. M. Knipe's *In the Image of fire*, 1975, pp. 90–121.

21 Ṛgv. VI.68.3d.

22 Cf. Ṛgv. V.85.7; II.28.9,5; VII.97.7; 88.7.

23 Ṛgv. II.28.5. The expression is used in J. Gonda, *Vision*, 1963, p. 114.

24 Ṛgv. I.24.9b; cf. Ṛgv. I.25.3. The use of the words *sumati* and *mṛlīka* is frequent: cf. Ṛgv. VII.86.2,7; 87.7; 88.1; 89.1; I.24.9; 25.3,5,19, etc.

Chapter 10 Forgiveness of sin and the meaning of freedom

1 Ṛgv. X.164.3; cf. X.12.5.

2 R. Panikkar, *The Vedic experience*, 1977, p. 483.

3 Cf. Ṛgv. I.23.22; 24.11,12; 25; II.28.9; 29.1–2,5; VII.86; 89; 93.7.

4 Panikkar, op. cit., p. 484.

5 A parallel case is found in the Greek term *metanoia* (cf. Luke 13:3,4,5 and II Corinthians 7:8,9) which means change of mind rather than repentance. Just as metaphysics means beyond purely physical science; or metamorphosis, a transformation of the structure of the grub into a completely different butterfly; so *metanoia* means beyond the present state of mind, a radical change of attitude. The Greek word does not

involve pain or sorrow as the English repentance does. Sorrow or repentance do not necessarily bring about a change of consciousness though they may lead towards it. But it is this change that is emphasised both in the Greek word and the Sanskrit, a radical turning over bringing about a complete change of attitude. Cf. R. Panikkar's view on repentance in *The Vedic experience*, 1977, p. 491.

6 *Ethics of India*, 1924, p. 33.

7 Cf. Ṛgv. I.67.2.

8 The many verbs used in connection with sin point this out: *ava-sṛj*, 'let loose, dismiss, pardon'; *śrath*, 'loosen, untie, remit pardon'; *vi-śrath*, 'loosen, untie, remit, pardon'; *pra-muc* 'release from, loosen, untie'. Cf. Psalms 103:12; II Samuel 12:13; John 1:29; Revelation 1:5.

9 Ṛgv. IV.12.4; cf. also Ṛgv. I.31.16; VII.93.7.

10 Ṛgv. IV.12.5ab. Cf. III.7.10cd and VI.51.8d: 'I desire-to-overcome even committed sin by homage.' The particle 'even' (*cid*) emphasising 'committed' (*kṛtam*) may infer that when a sin has been committed it is worse than when it remains at the intentional level, but it can still be pardoned.

11 Cf. Ṛgv. VII.89.5; VIII.47.13; I.185.8; X.35.3; VI.74.3 and 4; II.33.7; IV.54.3.

12 Cf. Ṛgv. I.162.22c; II.27.14; VII.93.7.

13 Ṛgv. VIII.67.7a; cf. X.63.4 and 35.3.

14 Ṛgv. II.27.14.

15 See B. G. Tilak, *The Arctic home in the Vedas*, 1903.

16 G. B. Tilak's trans. op. cit., p. 128. Tilak asks: 'Why should everyone be so anxious about safely reaching the other end of the night? And why should the poet exclaim that "its yonder boundary is not seen nor what keeps it separate". Was it because it was an ordinary winter night, or, was it because it was the long Arctic night?' (p. 128). One could ask why should the bard call this night 'O thou blessed one', according to Tilak's own translation. For a further assessment of the meaning of darkness in the Vedas see J. Miller, *The Vedas*, 1974, p. 162 ff.

17 J. Gonda offers the following explanation: 'It seems possible to take this line (*ā sarvatātim aditim vṛnīmahe*) as containing an explanation or qualification of sarvatāti; "we entreat a safe and sound condition", aditi- (freedom, free scope . . .) the wish: may we be, or remain, uninjured, is . . . extended by the prayer for 'free scope . . . unimpeded by distress, indigence" ' ('Gods and powers', 1957, pp. 78–9).

18 Cf. Ṛgv. I.185.3; IV.25.3; VII.82.10cd.

19 From J. Gonda's rendering of Ṛgv. X.185,2 and 3 (*Gods and powers*, p. 78).

20 *The Vedic idea of sin*, 1935, p. 16.

21 Ṛgv. II.33.3c.

22 Ṛgv. VII.65.3cd.

23 Ṛgv. V.4.9ab.

24 Ṛgv. I.106.1cd. Cf. also Ṛgv. X.133.6: 'Let us clasp Indra's friendship that by the path of *ṛta* it may guide us beyond all misfortunes' (*viśvāni duritā*).
25 Ṛgv. VIII.31.13c.
26 Ṛgv. III.59.2cd; cf. Ṛgv. X.133.6.

Chapter 11 *Ṛta* as truth: the Vedic vision of holiness, truth, and the social order

1 Panikkar, *The Vedic experience*, 1977, p. 483.
2 *The Rig-Veda and Vedic religion*, Christian Literature Society for India, 1913, p. 162.
3 See Rudolf Otto, *The idea of the Holy*, 1950, 2nd edn.
4 Cf. Ṛgv. VI.9.5cd.
5 Ṛgv. X.121.2.
6 Cf. Yjv. 19.30: 'By self-dedication one obtains consecration, by consecration grace, by grace reverence, and by reverence is truth obtained' (trans. A. C. Bose, *Hymns from the Vedas*, 1966, p. 59).
7 Ṛgv. X.90.1,2,3.
8 Ṛgv. I.24.6. Śri Aurobindo's translation with slight alteration. Cf. Ṛgv. VIII.41; Athv. IV.16.
9 Cf. Yjv. VS.40.8; Athv. IV.16.8.
10 Cf. Ṛgv. X.100, quoted in section on Forgiveness, chapter 10.
11 Ṛgv. IV.23,8b: *ṛtasya dhītir vṛjināni hanti*. Cf. Ṛgv. I.24.5; 25.1–3. Cf. also Psalm 119:97: 'O how love I thy law! it is my meditation all the day.'
12 *Ethics of India*, 1924, p. 44.
13 Panikkar, op. cit., p. 854.
14 Ṛgv. VII.62.2.
15 'The act of truth', *Journal of the Royal Asiatic Society*, 1917, p. 429.
16 'Duty as truth in the Ṛgveda', *India Maior*, ed. J. Ensink and P. Gaeffke, 1972, p. 59.
17 'The metaphysics of the Truth Act (*satyakriya*)', in *Mélanges d'Indianisme à la mémoire de L. Renou*, 1968, Institut de Civilisation Indienne, pub. fasc. 28, pp. 171–7.
18 Cf. Ṛgv. X.37.2.
19 For the whole subject see C. Kulkarni, *Vedic foundations of Indian culture*, 1973. See also R. Shamasastry, *Evolution of Indian polity*, 1920, for the history of Indian social order and the elective monarchy. For a discussion of the *sabhā* and *samiti* see F. B. J. Kuiper, 'The ancient Aryan verbal contest', in *Indo-Iranian journal*, vol. 4, no. 4, 1960, p. 264.
20 Ṛgv. VI.28.2, Dr Panikkar's trans., *Vedic experience*, p. 287.
21 'Duty as truth in the Ṛgveda,' p. 61.
22 Ṛgv. IX.112.1,3. Trans. slightly modified from A. C. Bose, *Hymns from the Vedas*, 1966, p. 197, 199.

23 This word has given rise to controversy. Geographically it means settlement; politically the *viśaḥ* were the subjects of a *rāṣṭra*: socially they represent the third class of Aryan society, the traders and agriculturists. Cf. Macdonell and Keith, *Vedic Index*, 1967 (1912), pp. 305–7. Cf. also J. Gonda, *Triads*, 1976, p. 138.

24 As noted by G. Dumézil, 'The sacred function is not presented in its agents but only in its principle and its means', i.e. the evocative power of the word or prayer, *brahman*. The same applies to the nobility here referred to under the notion of ruling power, *kṣatra*, which includes the idea of established sovereignty and physical force. For further elucidations see G. Dumézil, 'Les trois fonctions dans le Ṛgveda et les dieux indiens de Mitani', Académie de Belgique, Classe des lettres, Bulletin, 5e ser. 1961, pp. 265–98.

25 Ṛgv. I.111.4. Cf. also IV.50.8; V.54.14; 3.6; VIII.71.12. See also G. Dumézil, 'Mitre-Varuṇa, Indra, les Nāsatya comme patrons des trois fonctions cosmiques et sociales', *Studia linguistica*, 1947, no. 1, pp. 121–9.

26 *ṛtasya jyotiṣas patī* (I.23.5); cf. Ṛgv. V.68.1b: *mahikṣatrāvṛtam bṛhat*; VII.64.1: *kṣatriyā*; VIII.67.1; 25.8; I.136.3.

27 Ṛgv. IV.57.4–7, Dr Panikkar's translation (*Vedic experience*, p. 271). Cf. Athv. III.17.

28 Athv. III.30.1,2,4,6,7 (trans. A. C. Bose with slight alterations, *Hymns from the Vedas*, 1966, p. 155). Cf. also Ṛgv. X.25.1: 'Instil in us a wholesome, happy mind with good will and understanding. Then shall we ever delight in your friendship like cows who gladly rejoice in meadows green. This is my joyful message' (Dr Panikkar's translation, op. cit., p. 302). Cf. also Athv. I.31.4.

Chapter 12 The meaning of the Vedic sacrifice

1 *The call of the Vedas*, 1970, p. 49.

2 Panikkar, *The Vedic experience*, 1977, p. 348.

3 Ibid., p. 348.

4 Eggeling, *The Sacred Books of the East*, p. xv. Cf. New Testament, 1 Corinthians 15:36: 'the seed is not quickened except it die.'

5 Panikkar, op. cit., p. 75. Cf. also p. 347: 'the proper sphere of sacrifice is the sphere of communication, and communication constitutes the very structure of the universe.'

6 Judgments either so partial to other kinds of sacrifice that they were unable to perceive the characteristic of the Vedic rite (e.g. A. B. Keith), or plain incapacity to grasp the meaning of the Vedic sacrifice militated against any just appreciation of the subject by Western scholars at the turn of the century. Thus M. Bloomfield dismisses the whole subject as the 'foolish sacrifice' (*The religion of the Veda*, 1908, p. 76) or as sheer 'hocuspocus', (p. 75) and Winternitz shows Prajāpati, the later

equivalent of Puruṣa, preparing himself for the great work of creation through 'self-torture' and 'mortification' (*History of Indian literature*, vol. 1, 1927, p. 220, n. 2).

7 Ṛgv. X.90.1–4. We do not believe in human sacrifices being performed either in the time of the Ṛgveda or in that of the Brāhmaṇas. F. Kittel (*A tract on sacrifice*, Mangalore, 1872) quotes from the Śat. Br. to prove his point that human sacrifices were current; 'Man is indeed the first of the sacrificial victims' (Śat. Br. 6.2.1.18) as well as from the Atharvaveda: 'Thine (Oh Bhava) are these five victims, divided as cows, horses, men, goats and sheep' (Athv. XI.2.9). Understood figuratively it is the divine life in man (as well as animals) that is being sacrificed by being incarnated in matter. This interpretation tallies with another quotation from the Śat. Br.: 'This which is sacrificed is the soul of all beings and of all gods' (Śat. Br. 14.3.2.1), a quotation which confirms our main contention that the *puruṣa* or divine life through being imprisoned in the shackles of mortality or matter is the sacrificial victim *par excellence*. This sacrifice is necessary that the world might be a 'living' cosmos; man being, on earth, the 'king' of creation, whose divine nature is far more developed than the animals', is necessarily the 'first' among all the sacrificial victims. So the animal is used as a substitute for the human, 'the animal is man by reflection' or allegory (*pratimā*) (Śankhayana Śrauta Sūtra X.3; XI.8), and 'the animal is ultimately the sacrificer himself' (Ait. Br. II.11). Cf. Tait. Sam. VI.1.11.6; Śat. Br. 3.3.4.21.

8 *The religion of the Veda*, 1908, p. 65.

9 Śat. Br. 6.1.2.26.

10 In his article 'The sources and nature of *puruṣa* in the Puruṣa sūkta (Rigveda 10.90)', JAOS 51, 1931, W. N. Brown brings forward evidence to prove that 'Puruṣa has his chief importance as a blend of' the characteristics of Agni, Sūrya and Viṣṇu, and concludes: 'Puruṣa is both the essence of creatures and also the inclusive principle, the first principle, the ruler, the immortal, the eternal. . . . He does not . . . draw his importance from any relationship with the Teutonic world-giant or Ur-mensch, either genetically or ideologically; he is most significantly a secondary derivation from notions established antecedentally in the RV' (p. 114).

11 Ṛgv. VI.11.2d; X.81.5cd.

12 Cf. Bṛhad. Up. I.4.6; Taitt. Up. III.10.6; Aitareya Āran. II.1.2.15–16.

13 *The sacrifice in the Ṛgveda*, 1976, p. 143; p. 48.

14 Athv. XI.5.17 and 19. Cf. Ait. Br. II.13, where the gods are said to have conquered heaven by means of the sacrifice, austerities, penance and oblations. Also Mbh. XII.221.4–5: 'renunciation (*tyāga*) and humility are to be taught as the highest *tapas*; he who has got these two virtues, has also the continued fasting and persisting chastity.' John McKenzie (*Hindu ethics*, 1922) dismisses *tapas* in the Brāhmaṇas as calling 'for little special attention though it occupies a place of high importance'

(p. 32), thereby revealing both complete ignorance and conceit. The very meaning of *tapas*, which in man results in harmonising him with the cosmic order and thereby makes him 'holy' belies McKenzie's main contention that the ethics of Vedic literature are very meagre.

15 In this respect the following works are of great interest as showing contemporary attempts at fathoming the depth of meaning of *tapas* and not taking it merely at its literal or physical level as has been done hitherto: D. M. Knipe, *In the image of fire; Vedic experiences of heat*, 1975. This is so far the most complete study on the inner meaning of *tapas* in the Vedas. M. Eliade, *Yoga, immortality and freedom*, 1958, p. 106 ff. J. Gonda, *Gods and powers*, 1957, pp. 19, 32, 34, 94. Sri Krishana Prem, *The yoga of the Kathopanishad*, 1955, pp. 87, 88, 175.

Lama A. Govinda, *Foundations of Tibetan mysticism*, 1969. This is what is explained about *tapas* by one who knows its secret working from experience:

'The fire of spiritual integration which fuses all polarities, all mutually exclusive elements arising from the separateness of individuation, this is what the Tibetan word *g Tum-mo* means in the deepest sense and what makes it one of the most important subjects of meditation. It is the all-consuming incandescent power of that overwhelming Inner Fire which since Vedic times has pervaded the religious life of India: the power of *tapas*.

'*Tapas*, like *g Tum-mo*, is what arouses man from the slumber of worldly contentment, what tears him away from the routine of mundane life. It is the warmth of spiritual emotion which, if intensified, kindles the flame of inspiration, from which is born the power of renunciation and what appears to the outsider as asceticism. . . .

'*Tapas* is . . . the creative principle, which acts upon matter as well as upon mind. With regard to matter it is the forming, organizing, order-creating principle: "Out of the flaming *tapas* order and truth were born" (Ṛgv. 10.190.1). In the spiritual realm, however, it is that force which lifts us beyond the created, beyond that which has become, originated, taken form. It lifts us beyond the boundaries of our narrow individuality and of our self-created world. It dissolves and transforms all that which has "frozen" into rigid form' (pp. 160–1).

Again: 'The "Yoga of the Inner Fire" is therefore not concerned with the production of bodily heat – though this as well as a number of other extraordinary qualities may be created as a by-product. It is an often repeated misunderstanding to imagine that these practices were intended to enable the *Sādhaka* to survive in the icy mountain-solitudes of Tibet. Those who advance this theory forget that this yoga originated in the hot plains of India, where people would give anything to keep cool. The purpose of this yoga is therefore purely spiritual, aiming at a state of perfect inner unity and completeness in which all dormant forces and qualities of our being are concentrated and integrated like the rays of the sun in the focus of a lens. This process of perfect integration is represented by the symbol of the flame or the flaming drop. . . . The image of the flame is . . . not merely a metaphor, but the expression of real experiences of psycho-physical processes, in which all properties of fire, in their elementary (*tejas*) as well as in their subtle effects (*taijasa*) can make their appearance: warmth, heat, incandescence, purification and consummation by fire, fusion, upsurging flames, radiation, penetration, enlightenment, transfiguration, and so on' (p. 165).

16 Taitt. Br. 3.12.31 etc.; cf. also Athv. XII.1.39; VIII.10.25; XIX.43.1; 41.1; XVIII.2.15.

17 *Heat in the Rig Veda and Atharvaveda*, 1961, pp. 113, 114. Cf. as a contrast D. M. Knipe (op. cit.): 'Whether "head heat" derives from inspiration (*manīṣā*), from the cultivation of meditation techniques, or from the heat of Soma-intoxication and ecstasy, it is well known that the ṛṣi radiates a fiery "halo" like the *nimbus* of the Christian and Sūfi mystics' (p. 96). See also note 15 above.

18 From √*yuj* yielding yoga; cf. Ṛgv. V.81.1ab 'the inspired ones harness (*yuñjate*) their mind, harness their vision'.

19 *The Vedic experience*, 1977, p. 354.

20 Cf. Ṛgv. I.164.47.

21 *itthā dhiyā yajñavantaḥ á cakrur-agnim ūtaye* (Ṛgv. III.27.6); cf. V.1.7.

22 Cf. Ṛgv. X.101.2.

23 Ṛgv. IV.10.3b. The song of praise is like a milk cow that never fails. It draws the milk of heaven; cf. VIII.27.11.

24 Ṛgv. VIII.3.15bd.

25 *rāya ese'vase dadhīta dhīḥ* (V.41.5b). This is a case where *dhīḥ* may stand for the notion of revelation.

26 'Even to yonder vault has he outstretched it' (Ṛgv. X.130.2b).

27 Ṛgv. IX.73.9ab, in *Vision*, p. 212.

28 Cf. the Christian 'transubstantiation' of the bread and wine into the body and blood of the Lord.

29 *Vision*, p. 212.

30 E. W. Hopkins, *The religions of India*, 1895, p. 149.

31 Cf. L. Renou, *Religions of ancient India*, 1953, pp. 17–18, where he discusses how A. Bergaigne has shown us the right method, that of seeking correspondences between the world of men and that of gods, of matter and of spirit. J. M. Koller, 'Dharma: an expression of universal order', *Philosophy East and West*, vol. 22, 1972, pp. 140 and 135; H. Lefever, *The Vedic idea of sin*, 1935, pp. 2–3; and M. Biardeau and C. Malamoud, *Le sacrifice dans L'Inde ancienne*, École des Hautes Études, vol. 79, 1976, p. 20 ff. all bring out an important aspect of the sacrifice, namely the weaving of sacred links between creatures, especially those who are no longer united by visible links. This recognises 'the interdependence of all categories of beings and of all the parts of the universe'.

32 *Triads*, 1976, p. 49.

33 *Some aspects of the history of Hinduism*, 1967, p. 70.
34 *The sacrifice in the Ṛgveda*, 1976, p. 123.
35 *The religion and philosophy of the Vedas and the Upanishads*, p. 274. Cf. Buddha Prakash's views 'The meaning of yajña' in S.P.17th AIOC, Ahmedebad, 1953, p. 114. On the ground of comparison he maintains that yajña means 'eating' or 'feast'. M. Yamunacharya on the other hand maintains that yajña is worship in its best form ('The deeper meaning of yajña in Indian religious thought', *Quarterly Journal of the Mythic Society*, Bangalore, vol. 39, 1948, pp. 87–92).
36 For a complete study of Ṛgvedic sacrifice and a survey of scholars' contribution to the subject see K. R. Potdar, *Sacrifice in the Ṛgveda*, 1953.

H. Oldenberg's survey of animal sacrifices in the Ṛgveda is rather vague (*Die Religion des Veda*, 1970, pp. 354–66). Mention of animal sacrifice is made in Ṛgv. X.28.3; 27.2; 86.14; V.29.7; VI.17.11; the horse sacrifice in I.162 and 163. Cf. L. Renou, *Religions of ancient India*, 1953: 'We cannot . . . reconstruct this early cult. Our sole data would be the soma ceremonies which are the only rituals that the hymns treat in detail. . . . But the Ṛgveda, although it describes the preparation of the soma at great length hardly mentions the other operations. . . . Animal sacrifices for example, are barely touched upon' (p. 11). Cf. Deo-prakash Patanjal Shastri, *A critical study of Rigveda (1.137–163)*, 1963, pp. 34–36 and 386–423, where he examines Ṛgv. I.162 and 163 and argues in favour of a completely allegorical interpretation of these hymns and therefore of the horse sacrifice.
37 Ṛgv. VI.1.6cd; cf. also I.110.6
38 Ṛgv. I.128.2ab; cf. VII.51.1 where the sacrifice is stated to be a means to freedom and sinlessness.
39 Ṛgv. III.26.1ab; cf. II.35.12ab; I.93.8.
40 Ṛgv. I.68.2cd; cf. also the use of the word *cetana* in connection with Agni and the sacrifice (Ṛgv. III.3.8; II.5.1; I.170.4; VIII.13.18).
41 Cf. A. T. de Nicolas, *Four-dimensional man*, (1971): 'It is as a result of this transcendence (sacrifice) that Vedic man achieves in one auspicious intuitive moment that vision – identity of the Real which could make "his" action cosmically efficient' (p. 3).
42 *sadane ṛtasya* (Ṛgv. VII.53.2ab).
43 Cf. Ṛgv. I.139.1; VIII.32.16.
44 *Dhāman*, 1967, p. 31.
45 Ṛgv. VII.21.5d; I.133.1a; VII.101.6c; V.59.1b (cf. VII.38.2; V.21.4; III.12.7); X.8.5ab; I.185.10ab.
46 Ṛgv. X.111.2a and 138.2d.
47 Ṛgv. IV.2.16bc.
48 This is more explicit in the later literature. Cf. J. Gonda, *Triads*, 1976: 'The co-ordination of a ritual act and the cosmic fact of the threefold universe is at Ś.B. 6.5.2.22 explicitly based on an identity' (p. 57).
49 *dadhann ṛtaṃ dhanayann asya dhītim* (Ṛgv. I.71.3). J. Gonda explains this

verse thus: 'They established a special manifestation of ṛta and made . . . the intuitive and visionary "sight" of it an institution' (*Vision*, p. 174).

50 *pūrvā dhāmāny-amitā mimānāḥ* (Ṛgv. X.56.5b); *dhāma* would here be the institution of the rite.

51 Aguilar, *The sacrifice in the Ṛgveda*, 1976, p. 27.

52 Ibid., p. 38.

53 Ṛgv. X.181.2ab.

54 This will be examined in the section on Soma.

55 Literally 'charioteers' (Ṛgv. VIII.19.35cd: *syāma id ṛtasya rathyaḥ*).

56 Cf. J. McKenzie's contention that the sacrifice as understood and practised in Vedic times 'has no ethical significance' (*Hindu ethics*, 1922, p. 31). For an opposite view cf. A. C. Bose, *Hymns from the Vedas*, 1966, p. 42.

57 Ṛgv. X.13.3cd. For √*prati mā* cf. J. Gonda, *Four studies in the language of the Veda*, 1959, pp. 177–8.

58 Cf. Ṛgv. X.170.2; 85.1.

59 Ṛgv. IX.74.4b.

60 Ṛgv. IX.67.23abc; cf. VI.11.2.

61 *yad dīdayad divi* (Ṛgv. VI.16.36c).

62 Cf. also Ṛgv. I.44.10; 68.10; 72.1; VI.22.3; VIII.43.32; 46.8; IX.106.4; 107.21.

63 *trī sa pavitrā hṛdi antar ā dadhe* (Ṛgv. IX.73.8b).

64 Cf. Ṛgv. IX.97.34 and 111.2. Agni is also credited with three filters, cf. III.26.8.

65 *ṛtavākena satyena śraddhayā tapasā suta* (Ṛgv. IX.113.2c).

66 Cf. Ṛgv. I.67.2. This aspect will be considered in the last section of this chapter.

67 Cf. Ṛgv. IX.111.2 where Soma is said to be adored by the 'mothers' who seem to be the 'visions' (*dhiyaḥ*).

68 Ṛgv. III.12.7; cf. VIII.13.26b ('From the seat of *ṛta* I send thee a mind-yoked *dhīh*').

Chapter 13 The role of Soma and Agni in the ritual

1 Kramrisch, 'The triple structure of creation in the Ṛg Veda', *History of religions*, vol. 2, no. 1, summer 1962, pp. 143–4.

2 Cf. M. Bloomfield, 'The legends of Soma and the eagle', JAOS, vol. 16, 1896 pp. 1–24.

H. W. Magoun, 'Apām Napāt in the Rig-Veda', JAOS, vol. 19, pt 2, 1898, pp. 137 ff. Others, like Hillebrandt in *Vedische Mythologie* (Breslau, W. Koebner, 1891–1902), identified Soma with the moon even in the early part of the Ṛgveda where there is no such evidence, and Indian commentators never refer to it although during their time Soma was identified with the moon.

3 *The Rig Veda; an anthology*, 1981, p. 97.
4 Kramrisch, op. cit., p. 267.
5 Ibid., pp. 261, 265.
6 A *kaviḥ* is a poet in the ancient sense, i.e. a visionary sage who is inspired and inspires. Cf. L. Renou, *Religions of ancient India*, 1953, p. 10.
7 Ṛgv. IX.107.15.
8 Cf. I.77.1 where Agni is said to make the gods manifest (*kṛnoti devān*).
9 Ṛgv. IX.108.8cd; cf. IX.107.15 where Soma is also identified with *ṛta*.
10 Ṛgv. IX.56.1ab.
11 Ṛgv. X.177.2cd; IX.101.3.
12 Ṛgv. IX.95.5b; cf. J. Gonda's comment on *manīṣa*: 'like *dhīḥ manīṣa* is also the thought converted into a hymn or eulogy addressed or offered to a god' (*Vision*, p. 52).
13 *ātmā yajñasya* (Ṛgv. IX.6.8a).
14 *patir-dhiyaḥ* (Ṛgv. IX.75.2); cf. IX.99.6.
15 *patim vāco ādabhyam* (Ṛgv. IX.26.4c).
16 Cf. Ṛgv. I.139.2.
17 Ṛgv. IX.96.10d; cf. Agni, VII.13.3c.
18 Cf. M. Bloomfield, *The religion of the Veda*, 1908, p. 75.
19 Ṛgv. IX.97.34ab, translated in accordance with J. Gonda's comments (*Vision*, p. 52).
20 *Vision*, p. 178; cf. also p. 74: 'It may . . . be contended that the qualification "lord of *dhīḥ*" was given to Soma to characterise this divine power as being able or willing to dispense visions.'
21 Ibid., p. 178.
22 Ṛgv. IX.96.18; cf. Ṛgv. IX.111.2; 73.8b.
23 Here could be compared Ṛgv. I.164.45 where speech is said to be measured out in four sections of which three are kept concealed and only the fourth is known to men. Of these three, the first counting downwards would be equivalent to the fourth estate of Soma (counting upwards) that estate which is the most recondite. Cf. the third and fourth state of consciousness of the Maṇḍukya Upaniṣad.
24 *The Vedic experience*, 1977, pp. 98, 99.
25 Voice, reason and ratio are inseparable from sound. The rational mind creates order, hence relationship, out of the experience of the world, or the senses. Similarly sound has its own geometry, i.e. forms its own patterns in matter, as Chladni's plates experiment reveals.
26 Ṛgv. I.164.45cd, W. O'Flaherty, *The Rig Veda*, 1981.
27 Ṛgv. I.164.37c; cf. Śat. Br. 4.1.3.16; Taitt. Up. I.5.1–5 and Maitri Up. VI.6.
28 Ṛgv. IX.75.2 (*ṛtasya jihvā*); cf. Ṛgv. IV.58.1.
29 *Vision*, p. 74.
30 Ṛgv. IX.102.1b and 8c.

31 *manasas patiḥ* (Ṛgv. IX.11.8 and 28).
32 *kratur-indur-vicakṣana* (Ṛgv. IX.107.3b).
33 Cf. Ṛgv. IX.107.14; 21.1.
34 Ṛgv. I.139.2. This verse could also be translated: 'with the very eyes of Soma, indeed its very eyes', the meaning being the same, as 'its very eyes' become the eyes of the seer.
35 *urucakṣas* (Ṛgv. VII.35.8; 63.4).
36 Cf. Ṛgv. IX.110.3a,4ab; 10.8b.
37 Cf. Ṛgv. IX.75.1; 28.5; 85.9; 107.7; 2.6; 86.32a; 76.4c; 86.29.
38 Cf. Ṛgv. X.130.1a: 'the sacrifice drawn out with threads on every side'; Cf. IX.22.7d.
39 Ṛgv. VIII.13.14cd. To this verse Griffith adds the apposite note that 'the due performance of sacrifice is regarded as an unbroken thread through a succession of Rishis from ancient to modern times' (*Hymns of the Ṛgveda*, trans. R. T. H. Griffith, 4th edn., 1963, vol. 2, p. 136). This is quite in keeping with the ṛṣis' looking back to a golden age of seership whose continuity does not seem to have been broken. Cf. Ṛgv. I.109.3.
40 Cf. Ṛgv. IX.22.6; 82.3; 86.29.
41 *dharuṇo maho divoḥ* (Ṛgv. IX.72.7); cf. IX.86.8; 48.4.
42 Ṛgv. I.164.35. Although *brahmā* is in the masculine form and thus means the priest, the latter may be thought of in terms of prayer, he who wields the divine word, *brahma* (neuter), hence our translation emphasising the word rather than the wielder of the word.
43 Cf. Ṛgv. VIII.53.2; X.65.8; III.62.13 and 18.
44 Cf. Ṛgv. II.3.10; I.13.11.
45 Taitt. Sam. 6.6.4.2.
46 Śat. Br. 3.2.2.11.
47 Śat. Br. 6.2.1.18.
48 Murray Fowler, 'Old Norse religion' in *Forgotten religions*, edited by V. Ferm, New York, Philosophical Library, 1950, p. 242.
49 *The Poetic Edda*, Trans. H. A. Bellows, N.Y. Scandinavian Foundation, 1923.
50 *The sacrifice in the Ṛgveda*, 1976, pp. 143–4. For an excellent assessment of the idea of 'union and release' see the chapter 'The Vedic image of the world' (p. 125 ff).
51 Ṛgv. X.181.2ab and IX.74.4b.
52 Cf. R. Panikkar, *The Vedic experience*, 1977: 'In the performance of worship man always endeavors to transcend time. . . . By this liberation he enters into the sphere of ultimate reality. Liberation, mokṣa, is absolute freedom' (p. 354).
53 Cf. Ṛgv. X.45.5; IV.10.2; X.46.5; II.9.4d.
54 Ṛgv. IV.3.10; I.79.3.
55 *kavikratuḥ*.
56 Cf. Ṛgv. VI.7.4c; V.3.4; III.17.4; IX.106.8; VIII.48.3.
57 Cf. Ṛgv. VII.39.1; VI.15.14; I.94.13; VI.11.2b.

58 *The religion of the Rigveda*, 1923, pp. 164 and 166.

59 Cf. Ṛgv. III.10.2; VI.3.1; 13.2; VIII.60.5; III.2.10 (the people's 'inspirer' *viśāṃ kavim*). We prefer to translate *kavi* as 'inspirer' rather than 'poet', for the ancient meaning of poet is he who is inspired and capable of inspiring by his words, the modern meaning stressing too much the technicalities at the expense of the inspiration.

60 This very point is missed by V. A. Gadgil when he asks the question: 'How can Agni the most visible of all the Vedic gods be called a guardian of the Ṛta, the invisible eternal law? At the most he can guard Ṛta in some concrete form' ('Ṛta and the law of Karma', 10th Oriental Conference, Tirupati, 1940, p. 19). That concrete form being the expression, at the phenomenal level, of the universal law at the metaphysical level, there can be no difficulty in regarding Agni as the guardian of *ṛta* in its deepest sense.

61 Ṛgv. V.21.4d; I.75.5bc; III.10.2; I.1.8.

62 Ṛgv. VI.7.1; cf. I.44.12; 59.2; VIII.44.16; V.1.7c.

63 Ṛgv. III.11.2ab.

64 Ṛgv. I.143.7d; cf. I.144.1b.

65 Ṛgv. III.2.10c.

66 *adbhutasya rathīḥ* (Ṛgv. I.77.3).

67 Cf. Ṛgv. III.25.1ab.

68 Cf. Ṛgv. VI.2.10; 7.1.

69 Ṛgv. V.3.4c; cf. VI.1.1ab.

70 Ṛgv. VIII.19.4cd.

71 Ṛgv. III.27.11.

72 Ṛgv. III.5.3ab; 29.7 and 8.

73 *ṛtena vahā no havyam* (Ṛgv. X.12.2b).

74 Cf. Ṛgv. V.7.3; III.27.11.

75 *hotarām rodasyoḥ* (Ṛgv. VI.16.46).

76 Ṛgv. III.2.10. J. Gonda describes *tejas* as 'fiery energy, splendour, efficiency, majesty, supernormal potency, keenness' (*Gods and powers*, 1957, p. 58).

77 Cf. Ṛgv. V.51.2.

78 *ṛtāvṛdhaḥ* (Ṛgv. VII.66.10).

79 Cf. Ṛgv. I.89.7.

80 Ṛgv. IV.10.2; cf. I.128.4; III.27.12.

81 Cf. Ṛgv. IV.3.4b; I.127.9; *sukratuḥ* and *cikitvan* are epithets specifically applied to Agni. *Kratu*, a blend of purposeful will and intelligence, is defined by J. Gonda as: 'A kind of effective mental power or intelligence, mental energy and determination, which enables its possessor to have solutions for preponderantly practical difficulties' (*Vision*, p. 183). Cf. the same author's *Epithets in the Ṛgveda*, 1959, p. 37; and L. Renou, EVP., VII, p. 18, where commenting upon *kratu* in Ṛgv. V.85.2 the author writes that it 'seeṃs to mean "manas" but the heart being conceived as the seat of sacred inspiration the sense of "power of inspiration" can be maintained'.

82 *dhīmahi pracetasam* (Ṛgv. I.44.11).
83 Ṛgv. III.1.17b and 18d; cf. Ṛgv. X.21.5; II.5.3cd; IV.11.3ab.
84 *jātavedas; viśvavid; viśvavedas.*
85 *svarvid* (Ṛgv. III.26.1). This word could also be interpreted as 'knower of sound' and shows the close connection between Agni and Soma in their linking of sound and sight.
86 Ṛgv. IV.6.1; cf. I.140.1; VI.1.1.
87 *dhiye vājāya hinvatu* (Ṛgv. I.27.11). If *vāja* be a generative power, as J. Gonda explains (*Vision* p. 141), then this verse would mean that Agni helps to generate vision.
88 Ṛgv. IV.11.3ab.
89 Cf. Ṛgv. IX.66.20a where he is identified with Soma. Cf. also VI.14.2; III.21.3.
90 Cf. Ṛgv. IX.95.5; IV.11.2.
91 Cf. Ṛgv. I.144.1: 'The high priest performs his function, with his *māyā* he sends aloft his resplendent vision.' Cf. also I.143.7 and X.91.12 where *vāc*, *ṛc* and *gir* are offered up.
92 *atithir-gṛhe-gṛhe* (Ṛgv. X.91.2); cf. also X.46.7ab.
93 Ṛgv. X.91.14d; cf. VIII.43.11.
94 Cf. the Christian 'this is my body, this is my blood of the new covenant' (Matthew 26:26–28). There is here no shedding of blood, yet a direct assimilation with flesh and blood in a commemorative communion. Cf. Ṛgv. X.86.13 and 14; 27.2; 28.3; VI.17.11; V.29.7. One may wonder whether the references to Indra's devouring sacrificed animals may not also be figures of speech in like manner.
95 Cf. Ṛgv. VI.7.4; V.3.4.
96 *amṛtasya bhūrer-īśe* (Ṛgv. VII.4.6); *akṛnvan amṛtasya nābhim* (Ṛgv. III.17.4d); cf. IX.106.8. where the gods drink Soma for immortality.
97 *yajñasya dhāma paraman* (Ṛgv. X.181.2b).
98 √*bhūṣ*. Cf. J. Gonda's discussion of this verb in *Four studies in the language of the Veda*, 1959, pp. 75, 76, 92.
99 Cf. Ṛgv. III.15.4 and 5.
100 Cf. Ṛgv. I.31.7ab: 'Thou raisest the mortal, O Agni, to highest immortality.'
101 Ṛgv. V.85.2; cf. J. Gonda, *Epithets in the Ṛgveda*, 1959, p. 137.
102 The important part the heart plays in the Upaniṣads as the hidden cave wherein abides the *ātma*, the unborn, that which is the essence of the whole universe, can be traced back to this Ṛgvedic doctrine, a doctrine which stems from a vision.
103 Cf. Ṛgv. I.67.2; VII.33.9ab; V.3.3; IV.5.3.
104 Ṛgv. I.171.2a; VIII.76.8.
105 Cf. Ṛgv. X.63.7b.
106 *brahmaṇā vāvṛdhānā* (Ṛgv. I.93.6c).
107 Ṛgv. I.93.6cd.
108 *abrahmā yajña ṛdhag* (Ṛgv. X.105.8d); cf. VII.26.1; II.12.14.
109 Cf. Ṛgv. VII.26.1b.

110 Cf. Ṛgv. II.12.14; VI.38.4.
111 *sumnāyavaḥ* and *devayantaḥ*.
112 Ṛgv. VI.1.5 and 7ab.
113 Panikkar, *The Vedic experience*, 1977, p. 782.
114 *Hindu polytheism*, 1964, p. 68.
115 Panikkar, op. cit., pp. 28 and 351.

Chapter 14 *Ṛta*, world-embracing power

1 H. Lefever, *The Vedic idea of sin*, 1935, p. 6.
2 *Cosmology of the Rigveda*, 1887, p. 93.
3 A. B. Keith, *Religion and philosophy of the Veda*, 1925, p. 84.
4 Cf. Ṛgv. IV.23.10; V.12.2; I.75.5; IV.3.9.
5 Cf. Ṛgv. X.129.1 and 7, and A. Daniélou's discussion of 'The representation of the Transcendent' in *Hindu polytheism*, 1964, pp. 4–5.
6 Ṛgv. X.151.5; cf. IV.57 addressed to the lord of the field, to the plough and the furrow; I.188.8; 31.11; VII.44.2 addressed to Iḷā first personified, then deified.
7 Keith, op. cit., p. 210. Cf. J. Gonda, *Gods and powers*, 1957, pp. 19 and 36. In the Atharvaveda VI.41 'worship is paid with oblation to *manas* "mind", *cetas* "thought", *cakṣas* "visual faculty" and other powers' (p. 36).
8 Ṛgv. X.190.1.
9 Cf. Ṛgv. VIII.89.4.
10 E.g. *dhāman ṛtasya* (I.43.9; IV.7.7); *dharman ṛtasya* (IX.7.1).
11 E.g. *sadas* (X.111.2; III.7.2); *raśmi* (V.7.3; I.123.13); *cakram* (I.164.11); cf. *ṛtasya dhīti* (I.68.5; IX.111.2); *ṛtasya garbham* (I.156.3; IX.68.5); *ṛtasya jihva* (IX.75.2); *ṛtasya pada* (X.177.2; IV.5.9; X.5.2); *ṛtasya yoni* (X.68.4; 65.7; III.54.6); *ṛtasya pantha* (V.80.4; X.31.2; 133.6; I.136.2; 46.11; IX.76.4, etc.). All these may indicate something more than 'the natural concreteness of expression of the Vedic age' (Keith, op. cit., p. 84). In K. R. Potdar's opinion 'From the variety of epithets derived from "Ṛta", applied to the gods as "*ṛtaprajāta*", "*ṛtajña*", "*ṛtadyumnā*", "*ṛtapeśaḥ*", "*ṛtāvānaḥ*", "*ṛtāvṛdhaḥ*", etc., it is obvious that at one stage it was being considered as the all-pervading and the all-supporting principle. In the later days however, only the sacrificial aspect of the Ṛta seems to have survived' ('Sacrificial setting of the philosophical hymns in the Ṛgveda', in *Bharatiya Vidya*, 12, 1951, p. 171).

For V. M. Apte the fact that the dawn is described as coming from the place of *ṛta* shows that *ṛta* is a concrete entity with a specific location in the heavens ('The Ṛgvedic antecedents of the dharma-pāśa of Varuṇa in the Mahābhārata', *Deccan College Research Institute Bulletin*, vol. 5, 1943–4, pp. 185–8).

For H. Lefever this is one among other 'mythological references' which, not being found in connection with *yajña*, the sacrifice, and other law terms, such as *dharma, dhāman, vrata,* indicate 'that ṛta was different in kind from the conception of law expressed by the other terms' (op. cit., p. 6).

12 Cf. Ṛgv. II.28.4; IV.56.7; V.62.1; IX.66.24 where Soma engenders *ṛta.*

13 *The Vedic idea of sin,* 1935, pp. 9 and 13.

14 Cf. Ṛgv. I.151.4bcd; IV.23.10; I.156.5; II.28.4; IV.56.6 and 7.

15 Cf. Ṛgv. IX.108.8; V.69.1.

16 V. M. Apte, 'The Ṛgvedic antecedents of the dharma-pāśa of Varuṇa in the Mahābhārata', *Deccan College Research Institute Bulletin,* vol. 5, 1943–4, p. 181.

17 Ṛgv. I.137.2; cf. IV.23.8–10.

18 *Ethics of India,* 1924, pp. 40–1.

19 The *ṛta* of Varuṇa as already explained does not necessarily refer to the source of *ṛta.* That source is in the original act of creation, *tapas* from the impersonal *tat,* 'That' (Ṛgv. X.190.1; X.129.3). All the devas are offsprings of *ṛta* (*ṛtajātaḥ*) and the first born of *ṛta* (*prathamajā ṛtasya*). This cannot be otherwise as *ṛta* is the first to manifest (X.190.1).

20 *The hymns of the Ṛgveda,* 1963, vol. 1, p. 190.

21 *La religion védique,* 1883, p. 231.

22 V. M. Apte criticises Oldenberg for his translation: 'Sacrifice to the gods (a sacrifice conforming to) the great ṛta', and Geldner's 'Adore the lofty law (Gesetz) of the gods' as due 'to the non-recognition of the divine status – the godly rank – of ṛta which entitled it to the offering of Soma and sacrificial worship generally' (op. cit., p. 182). This verse was translated thus by R. T. H. Griffith: 'Bring to us Mitra Varuṇa, bring the gods to mighty sacrifice, bring them Agni to thine own home'. √*yaj* means to 'worship' with sacrificial offering rather than 'bring', although the very act of worshipping effects a bringing together of heavenly and earthly powers. Nevertheless here again we have another example of a strict parallel case, where *devān* is the object of *yajā* and *ṛtaṃ bṛhat* is also the object of *yajā.* H. Lefever sees in the epithet *bṛhat,* so often used with *ṛta,* an example of an incipient personification (op. cit., p. 8). This 'worship with sacrifice' is thus, strictly speaking, addressed to Mitra, Varuṇa, to the gods, to the Cosmic Order.

23 *The call of the Vedas,* 1970, p. 49.

24 Ṛgv. V.12.2, J. Gonda's translation (*Vision,* p. 173).

25 Ṛgv. IX.74.4.

Chapter 15 Conclusion: *ṛta,* cosmic harmony

1 Cf. M. Winternitz, *A history of Indian literature,* 1927, vol. 1; A. L. Basham, *The wonder that was India,* 1954 (reprinted 1974), pp. 234–43; A. C. Bouquet, *Hinduism,* 1966, p. 44; R. C. Zaehner, *Hinduism,* 1966,

ch. 1; S. Radhakrishnan and C. Moore, *A source book in Indian philosophy*, 1967, p. 3.

2 Radhakrishnan, op. cit., p. 3.

3 *Four-dimensional man*, 1971, p. 151.

4 Ṛgv. VIII.58.2d: *ekam vā idam vi babhūva sarvam*; cf. also X.114.5; I.164.46.

5 Ṛgv. III.54.8cd.

6 *Some aspects of the history of Hinduism*, 1967, p. 32.

7 Katha Up. I.2.23; cf. Muṇḍ. Up. 3.2.3.

8 Ṛgv. I.164.20; cf. with Muṇḍ. Up. 3.1.1 and Bhagavad Gita 15:16, Katha Up. 3.1 and 6.1, and see article by Willard Johnson for a complete survey of scholars' contribution to the interpretation of the verse – 'On the Ṛgvedic riddle of the two birds in the fig tree', JAOS, vol. 96, Apr.-June 1976, pp. 248–58.

9 Cf. Vaj. Sam. 10.24; 12.14; Taitt. S. 3.2.10.1; Śat. Br. 6.7.3.11; Maitri Up. 6.4.

10 See Śvet. Up. 3.3. and cf. Ṛgv. X.72.2; X.81.3; Vaj. Sam. 17.19; Athv. 13.2.26; Taitt. S. 4.6.2.4. etc.; Mahan. Up. 1–14.

11 Cf. Ṛgv. I.24.9d; III.7.10d; VII.86.3a; etc. Cf. also I.89.8 with the opening lines of the Praśna Up. and Muṇḍaka Up.

12 *The Vedic experience*, 1977, pp. 315–16.

13 Cf. Śat. Br. 14.3.21: 'That which is sacrificed is the soul of all beings and all gods'. Cf. also Ṛgv. X.81.5 and 6.

14 *svayuṃ yajasva tanvaṃ vṛdhānaḥ* (Ṛgv. X.81.5d).

15 Cf. L. Renou, *Religion of ancient India*, 1953: 'The Vedic and the Upaniṣadic texts both seek the same end but they use different means' (p. 18). Cf. M. Winternitz, *History of Indian literature*, 1927, vol. 1, p. 225.

16 Cf. Bhagavad Gita: 'Better one's own duty though destitute of merit, than the duty of another, however well discharged' (III.35; cf. XVIII.48).

17 Māhān. Up. 62.6; 63.7.

18 References to the four *yugas* or ages of the world and to a 'day' of Brāhmaṇa, however, are found in the Devatādhyāya, chapter 3. Cf. also Ṛgv. X.97.1; 72.2, as against J. Muir's statement that 'Of this elaborate system of Yugas, Manvantaras, and Kalpas, of enormous duration, no traces are found in the hymns of the Rig-veda'. (OST IV, p. 45). He then proceeds to examine all the Ṛgvedic verses where the word *yuga* appears, an examination which only helps to convince one that the doctrine of *yugas* must have been known, the details of which may have been kept secret.

19 Cf. Ṛgv. X.190.3; 97.1; IV.54.2; Śat. Br. 7.2.4.26: 'The herbs first grown three ages before the gods'.

20 Cf. J. N. Farquhar, *An outline of the religious literature of India*, 1920 (reprinted 1967): 'There is no trace of transmigration in the hymns of the Vedas' (p. 33). Cf. A. B. Keith, 'Pythagoras and the doctrine of transmigration', JRAS, 1909, where on pp. 574–5 he discusses the

Hindu doctrine and states that 'most authorities are agreed that it can be found only in the Upaniṣads'. Hopkins, *The Religions of India*, ascribes it first to the Śat. Br. with a possibility in Ṛgv. I.164.30,38. Cf. A. A. Macdonell, *History of Sanskrit literature*, p. 115; Geldner, *Vedische Studien*, II.288; III.3. The discussion centres mainly around Ṛgv. I.164.38, which verses are most ambiguous, and no account is taken of Ṛgv. IV.54.2, which is the only clear statement in the Ṛgveda, but which A. A. Macdonell renders as 'length of life on man' (*Vedic mythology*, 1897, p. 32). Cf. K. Werner, 'The Vedic concept of human personality and its destiny', *Journal of Indian Philosophy*, 5, 1978, p. 287, for a counter argument. Cf. R. C. Zaehner, *Hinduism*, p. 57: 'Of this [the doctrine of rebirth, reincarnation, of the transmigration of souls] there is no trace in the Samhitās or the Brāhmaṇas.'

21 Cf. Śat. Br. 3.1.2.21; cf. also the beginnings of the idea of *devayāna* and *pitṛyana* in the Ṛgveda 1.162.4; IV.37.1; X.2.7; 88.15; 181.3.

22 Bṛh. Up. 3.2.13.

23 *satya* from *sat*, what is, what essentially is, in strict accordance with its own essence. Cf. H. Lefever, *The Vedic idea of sin*: 'This term (*satya*) . . . was admirably suited to express the conception of "right", "truth", or "law" in a system where Reality was equated with something spoken, i.e. brahman, the prayer, or with the power which was inherent in the prayer (cf. Śat. Br. 2.1.4.10)' (p. 57).

24 Śat. Br. 9.5.1.18.

25 Cf. Śat. Br. 7.4.1.8: 'This earth is established on the truth; hence the truth is this earth' with Ṛgv. X.85.1 and 1.105.12b.

26 Cf. Ait. Br. 3.1.6.(3): 'The sky rests on the air; the air on the earth; the earth on the waters; the waters on the reality (*satya*); the reality on the brahma; the brahma on the concentrated heat of meditation (*tapas*).' Cf. Chānd. Up. VIII.4.3.: 'And that world of brahman belongs to those only who find it by abstinence.'

27 Śat. Br. 2.2.2.20; cf. also Śat. Br. 9.5.1.18: 'Having made up the *yajña* let us spread out this *satya*.'

28 Śat. Br. 2.1.4.10.

29 Chānd. Up. VIII.3.4.

30 Bṛhad. Up. V.5.1; cf. V.4.1.

31 Mahānā. Up. 1.6; cf. XII.1.

32 Tait. Up. II.1.

Bibliography

This is only a select list of books and articles on the subject. Abbreviations used in the text are placed at the end of the relevant entries.

Original sources

Vedas

Ṛgveda Saṃhitā, ed. Max Müller, 4 vols, 2nd edn. London, 1890–92. Trans. R. T. H. Griffith, Varanasi, 1963 (reprint of 1899) (Ṛgv.).
Atharvaveda Saṃhitā, ed. R. Roth and W. D. Whitney, Berlin, 1955. Trans. W. D. Whitney, Cambridge, Mass., 1905 (H.O.S. vols 7 and 8) (Athv.).
Yajurveda, Vājasaneyī saṃhitā (VS), ed. A. Weber, London, 1852. Texts of the White Yajurveda, trans. R. T. H. Griffith, Benares, 1889 (Yjv.).
Taittirīya Saṃhitā, Calcutta, Bibliotheca Indica, 1860. Trans. A. B. Keith, Cambridge, Mass., Harvard Oriental Series, vol. 19, 1914.

Brāhmaṇas

Aitareya Brāhmaṇa of the Rgveda, ed., trans. and explained by M. Haug, Allahabad, 1974, (reprint of 1922) (Ait. Br.).
Śatapatha Brāhmaṇa, Sacred Books of the East, trans. J. Eggeling, Oxford, Clarendon Press, 1882, (Śat. Br.).
Taittirīya Brāhmaṇa, Poona, 1898 (Tait. Br.).

Upaniṣads

Bṛhadāraṇyaka Upaniṣad, *Thirteen principal upaniṣads*, R. E. Hume, Oxford, 1962 (reprint of 1921) (Bṛh. Up).
Chāndogya Upaniṣad, Bibliotheca Indica, 1849–50 (Chānd. Up.).
Mahānārāyaṇa Upaniṣad, ed. G. A. Jacop, Bombay, 1888 (Mahānā. Up.).
Śvetāśvatara Upaniṣad, Thirteen principal upaniṣads, R. E. Hume, Oxford, 1962 (Śvet. Up.)
Taittirīya Upaniṣad, Eight Upaniṣads, with the commentary of Śaṇkarācārya, trans. Swami Gambhīrananda, Calcutta, 2nd edn. 1965 (Tait. Up.)

Kalpa Sūtras

Śāṇkhāyana Śrauta Sūtra, Calcutta, 1886–9.

Critical works

Aguilar, H., *The sacrifice in the Ṛgveda*, Delhi, Bharatiya Vidya Prakasham, 1976.
Apte, V. M., 'All about "vrata" in the Ṛgveda', *Deccan College Research Institute Bulletin*, Poona, vol. 3, no. 4, 1942, pp. 407–88.
Apte, V. M., 'Is Diti in the Ṛgveda a mere reflex of Aditi?' *Bharatya vidya*, Bombay, vol. 9, 1949.
Apte, V. M., 'The Ṛgvedic antecedents of the dharma-pāśa of Varuṇa in the Mahābhārata', *Deccan College Research Institute Bulletin*, Poona, vol. 5, 1943–4.
Apte, V. M., 'Ṛta in the Ṛgveda', *Bhandarkar Oriental Research Institute Annals*, Poona, Silver Jubilee volume 1917–42, vol. 23, 1942 (BOR Institute).
Aurobindo, *On the Veda*, Pondicherry, Shri Aurobindo Ashram, 1964.
Barth, A., *Quarante ans d'Indianisme*, Oeuvres, tome 1, in the series *Les religions de l'Inde*, Paris, Ernest Leroux, 1914.
Basham, A. L., *The Wonder that was India*, London, Collins, 1954 (reprinted 1974).
Benveniste, E. and Renou, L., 'Vṛtra et Vreragna: Étude de mythologie indo-iranienne', *Cahiers de las Société Asiatique*, Paris, 1934.
Bergaigne, A., *La religion védique*, Paris, École des Hautes Études, 1878–83, 3 tomes.
Bhattacharji, S., *The Indian theogony: A comparative study of Indian mythology from the Vedas to the Purāṇas*, Cambridge University Press, 1970.
Biardeau, M. and Malamoud, C., *Le sacrifice dans l'Inde ancienne*, Paris, Presses Universitaires de France, 1976 (École des Hautes Études Bibliothèque, vol. 79).

Blair, C. J., *Heat in the Rig Veda and Atharva Veda*, New Haven, American Oriental Society, 1961 (American Oriental series, vol. 45).

Bloomfield, M., *The religion of the Veda*, New York, London, Putnam's Sons, 1908. 'The legend of Soma and the eagle', *Journal of the American Oriental Society*, vol. 16, 1896 (*JAOS*).

Boehtlingk, O. and Roth, R., *Sanskrit-Wörtebuch*, St Petersburg, Kaiserlichen Akademie der Wissenschaften, 1855–75.

Bose, A. C., *The call of the Vedas*, Bombay, Bharatiya Vidya Bhavan, 1970. *Hymns from the Vedas*, London, Asia Publishing House, 1966.

Bouquet, A. C., *Hinduism*, London, Hutchinson University Library, 1966.

Brough, J., 'The tripartite ideology of the Indo-Europeans: an experiment in method', *School of Oriental and African Studies Bulletin*, vol. 22, pt 1, 1959.

Brown, W. N., 'The creation myth of the Rig Veda', *Journal of the American Oriental Society*, vol. 62, 1942, pp. 85–98 (*JAOS*).

Brown, W. N., 'Duty as truth in the Rig Veda', *India Maior*, ed. J. Ensink and P. Gaeffke, Leiden, Brill, 1972, pp. 57–67.

Brown, W. N., 'Proselytising the Asuras; a note on Rig Veda X.124', *Journal of the American Oriental Society*, vol. 39, 1919, pp. 100–103 (*JAOS*).

Brown, W. N., 'The Rig Vedic equivalent for hell', *Journal of the American Oriental Society*, vol. 61, 1941, pp. 76–80 (*JAOS*).

Brown, W. N., 'The sources and nature of puruṣa in the Puruṣa sūkta (Rigveda 10.90)', *Journal of the American Oriental Society*, vol. 51, 1931, pp. 108–18 (JAOS). 'Theories of creation in the Rig Veda', JAOS, vol. 85, no. 1, 1965.

Buddha Prakash, 'The meaning of yajña', S.P. 17th All India Oriental Congress, Ahmedabad, 1953 (AIOC).

Cambridge History of India, vol. 1, 'Ancient India', ed. E. J. Rapson, Delhi, S. Chand & Co., 1968 (reprint).

Clayton, A. C., *The Rig-Veda and Vedic religion, with readings from the Vedas*, London, Christian Literature Society for India, 1913.

Coomaraswamy, A. K., 'Angel and Titan: an essay in Vedic ontology', *Journal of the American Oriental Society*, vol. 55, 1935 (*JAOS*).

Coomaraswamy, A. K., *The Vedas: essay in translation and exegesis*, Beckenham, Prologos Books, 1976.

Cunningham, Sir A., *Book of Indian eras*, Calcutta, Thacker Spink & Co., 1883.

Dandekar, R. N., 'Asura Varuṇa', *Bhandarkar Oriental Research Institute Annals*, Poona, vol. 21, Apr.–July 1940 (BOR Institute).

Dandekar, R. N., *Some aspects of the history of Hinduism*, Poona, University of Poona, 1967.

Dandekar, R. N., 'Universe in Vedic thought', *India Maior*, Leiden, Brill, 1972.

Daniélou, A., *Hindu polytheism*, London, Routledge & Kegan Paul, 1964.

Deoprakash Patanjal Shastri, *A critical study of Rigveda (1.137–163) particularly from the point of view of Paninian grammar*, New Delhi, Patanjal Publications, 1963.

Deshmukh, P. S., *The origin and development of religion in Vedic literature*, London, Oxford University Press, 1933 (*Religion in Vedic literature*).

Dumézil, G., *Les Dieux des Indo-Européens*, Paris, Presses Universitaires, 1952.

Dumézil, G., *L'idéologie tripartie des Indo-Européens*, Bruxelles, Bercham, 1958 (Collection Latomus, vol. 31, pp. 5–122).

Dumézil, G., *Mitra Varuṇa: Essai sur deux représentations indo-européennes de la souveraineté*, Paris, Gallimard, 1948.

Dumézil, G., 'Ordre, fantaisie, changement dans les pensées archaiques de l'Inde et de Rome', *Revue des études latines*, tome 32, 1954, Paris ('Ordre, fantaisie').

Eliade, M., *Yoga, immortality and freedom*, London, Routledge & Kegan Paul, 1958.

Farquhar, J. N., *An outline of the religious literature of India*, Delhi, Motilal Banarsidass, 1967.

Gadgill, V. A., 'Ṛta and the law of karma', 10th All India Oriental Conference, Tirupati, 1940 (AIOC).

Gauchwal, B. S., 'The metaphysical foundations of Hindu ethics and religion', *Philosophy East and West*, Honolulu, vol. 16, July–Oct. 1966.

Geldner, K. F., *Vedische Studien: Der Rigveda*, Cambridge, Mass., Harvard University Press, 1951.

Gonda, J., *Epithets in the Ṛgveda*, 'S-Gravenhage, Mouton, 1959.

Gonda, J., *Four studies in the language of the Veda*, 'S-Gravenhage, Mouton, 1959 (*Four studies*).

Gonda, J., *Loka: World and heaven in the Veda*, Amsterdam, Noord-Hollandsche Uitgevers Meatschappij, 1966 (*Loka*).

Gonda, J., *The meaning of the Sanskrit term dhāman*, Amsterdam, Noord-Hollandsche Uitgevers Meatschappij, 1967 (*Dhāman*).

Gonda, J., *Some observations on the relations between 'Gods' and 'powers' in the Veda à propos of the phrase sūnuḥ sahasaḥ*, 'S-Gravenhage, Mouton, 1957 (*Gods and Powers*).

Gonda, J., *Triads in the Veda*, Amsterdam, North Holland Pub, Co., 1976 (*Triads*). 'The Vedic concept of aṃhas', *Indo-Iranian Journal*, 'S-Gravenhage, Mouton, vol. 1, 1957, pp. 33–60.

Gonda, J., *Vedic literature, History of Indian literature*, vol. 1, Wiesbaden, Otto Harrasowitz, 1975.

Gonda, J., *The Vision of the Vedic poets*, The Hague, Mouton, 1963 (*Vision*).

Govinda, Lama A., *Foundations of Tibetan mysticism*, London, Rider, 1960.

Grassmann, H., *Wörterbuch zum Rig-veda*, Leipzig, F. A. Srockhaus, 1873.

Griswold, H. D., *The religion of the Rigveda*, London, Humphrey Milford, 1923. *The God Varuṇa in the Rigveda*, New York, Taylor and Carpenter, 1910.

Hiryanna, M., *Outlines of Indian philosophy*, Bombay, Allen & Unwin, 1973.

Hopkins, E. W., *Ethics of India*, New Haven, Yale University Press, 1924.

Hopkins, E. W., *The religions of India*, Boston, Ginn & Co., 1895.

Hopkins, E. W., 'Theories of sacrifice as applied to the Ṛgveda', *Journal of the American Oriental Society*, vol. 16, 1896 (*JAOS*).

Hubert, H. and Mauss, M., *Sacrifice: Its nature and function*, trans. by W. D. Halls, London, Cohen & West, 1964.

India Maior, Congratulatory volume presented to J. Gonda, ed. by J. Ensink and P. Gaeffke, Leiden, Brill, 1972.

Keith, A. B., *Indian mythology*, Boston, Marshall Jones, 1917.

Keith, A. B., *The religion and philosophy of the Veda and the Upanishads*, Cambridge, Mass., 1925 (*Religion and philosophy of the Veda*).

Kittel, F., *A tract on sacrifice (yajña-sudhanidhi)*, Mangalore, C. Stolz, 1872.

Knipe, D. M., *In the image of fire, Vedic experiences of heat*, Delhi, Motilal Banarsidass, 1975.

Koller, J. M., 'Dharma: an expression of universal order', *Philosophy East and West*, Honolulu, vol. 22, Apr. 1972.

Kramrisch, S., 'The Triple structure of creation in the Ṛgveda', *History of Religion*, vol. 2, no. 1, 1962 (*HR*).

Kristensen, W. B., *The meaning of religion: Lectures in the phenomenology of religion*, trans. by J. B. Carman, The Hague, Martinus Nijhoff, 1960.

Kuiper, F. B. J., 'The ancient Aryan verbal contest', *Indo-Iranian Journal*, vol. 4, no. 4, 1960, pp. 218–81 (*IIJ*).

Kulkarni, C., *Vedic foundations of Indian culture*, Bombay, Shri Dvalpayana Trust, 1973.

Lefever, H., *The Vedic idea of sin*, Nagercoil, London Missionary Press, 1935.

Linden, C. W. J. van der, *The concept of deva in the Vedic age*, Utrecht, Vesta, 1954.

Lüders, H., *Varuṇa*, Göttingen, Vandenhoeck & Ruprecht, 1951, 1953.

Macdonald, K. S., *The Brāhmaṇas of the Veda*, London and Madras, Christian Literature Society for India, 1901.

MacDonald, K. S., *The Vedic religion*, London, Nisbet, 1881.

Macdonell, A. A., *Vedic grammar*, *Grundriss der Indo-Arischen Philologie und Altertumskunde*, Band 1, heft 4, Strassburg, Trübner, 1910.

Macdonell, A. A., *Vedic mythology*, *Grundriss* etc., Band 111, heft A, Strassburg, Trübner, 1897.

MacKenzie, J., *Hindu ethics*, London, Oxford University Press, 1922.

Miller, J., *The Vedas: harmony, meditation, fulfilment*, London, Rider, 1974.

Miller, J. and Feuerstein, G., *A reappraisal of yoga*, London, Rider, 1971.

Muir, J., *Original Sanskrit texts*, London, Trübner, 1858–70 (OST).

Müller, Max, *Lectures on the origin and growth of religion*, London, Longmans, Green, 1878.

Muller, R. F. G., 'Antarikṣa, ākāśa, kha', *Rocznik orientalistyczny*, Krakow, tome 24, pt 2, 1961, pp. 53–8.

Narahari, H. G., 'Designation of hell in the Ṛgveda and the meaning of the word "asat" ', *Indian historical quarterly*, Calcutta, vol. 18, June 1942.

Neisser, W., *Beiträge zur Sprach- und Völkerkunde: Festschrift für den Geheimen Regierungsrat*, A. Hillebrandt, Haile, Buchhandlung des Waisenhauses, 1913.

Nicolas, A. T. de, *Four-dimensional man: the philosophical methodology of the Rigveda*, Bangalore, Dharmaram College, 1971.

The Nighaṇṭu and the Nirukta, ed. Lakshman Sarup, Delhi, Motilal Banarsidass, 1967.

Oertel, H., 'Asat = undifferentiated, formless, incapable of perception by the senses in Vedic prose', *New Indian antiquary*, Bombay, vol. 1, no. 5, 1938, pp. 317–21.

O'Flaherty, W. D., *The Rig Veda; an anthology*, London, Penguin, 1981.

Oldenberg, H., *Die Hymnen des Rigveda*, Berlin, Wilhelm Hertz, 1888. *Die Religion des Veda*, Berlin, Wilhelm Hertz, 1894 (reprinted 1970).

Oldenberg, H., 'Zur Religion und Mythologie des Veda', *Nachrichten von der Königlichen Gesellschaft der Wissenschaften zu Göttingen. Philologische Historische Klasse*, Berlin, Weidmannsche Buchhandlung, 1916.

Otto, R., *The idea of the holy*, trans. J. W. Harvey, London, Oxford University Press, 1950.

Panikkar, R., *The Vedic experience*, London, Darton, Longman & Todd, 1977.

Parab, B. A., *The miraculous and mysterious in Vedic literature*, Bombay, Popular Book Depot, 1952.

Pischel, R. and Geldner, K. F., *Vedische Studien*, Stuttgart, W. Kohlhammer, 1889–1901.

Potdar, K. R., *Sacrifice in the Ṛgveda*, Bombay, Bharatiya Vidya Bhavan, 1953. 'Sacrificial setting of the philosophical hymns in the Ṛgveda', Bombay, *Bharatiya Vidya*, vol. 12, 1951, pp. 163–71.

Radhakrishnan, S. and Moore, C., *A source book of Indian philosophy*, Princeton, Princeton University Press, 1967.

Ragozin, Z. A., *Vedic India*, Delhi, Munshi Ram Manohar Lal, 1961, 2nd edn.

Rajnade, V. K., 'Asurasya māyā in Ṛgveda', *First Oriental Conference*, Poona, procs. and trans., 1920, vol. 1.

Ranade, R. D., *A constructive survey of Upanishadic philosophy*, Bombay, Bharatiya Vidya Bhavan, 1968.

Renou, L., *Études védiques et paninéennes*, Paris, Publication de l'Institut de civilisation indienne, 1955 (*EVP*).

Renou, L., 'Les origines de la notion de "māyā" dans la spéculation indienne', *Journal de psychologie normale et pathologique*, Paris, vol. 41, 1948.

Renou, L., 'Un hymne à énigmes du Ṛgveda', *Journal de psychologie*, Paris, vol. 42, 1949.

Renou, L., *Religions of ancient India*, London, Athlone Press, 1953.

Renou, L., 'Védique nirṛti', *Indian linguistics*, Lahore, vol. 16, Nov. 1955.

Roth, R., *Abhandlung über der Atharva Veda*, Tübinger Universitätsschriften, 1856. *Die höchsten Götter der arishcen Völker*, Z.D.M. Gessellschaft, Leipzig, 1852.

Shamasastry, R., *The conception of sin in the Vedas: Festschrift Moriz Winternitz (1863–1933)*, Leipzig, Otto Harrassowitz, 1933.

Siqueira, T. N., 'Sin and salvation in early Rigveda', *Antrhopos*, Salzburg, vol. 28, 1933, p. 185.

Tavadia, J. C., 'Meaning of ṛta', *Bhandarkar Oriental Research Institute Annals*, Poona, vol. 35, 1954.

Thieme, P., 'Beitrage zur Vedaexegese. 1. Die beiden Verben mī', *Zeitschrift der Deutschen Morgenländischen Gesellschaft*, 95, 1941, pp. 82–116 (ZDMG).

Tilak (Bāḷa Gaṅgādhara Ṭiḷaka), *The arctic home in the Vedas*, Poona, 1903.

Varenne, J. *Cosmogonies védiques*, Milano, Archè, 1982.

Varma, K. S., 'The doctrine of māyā', *Agra University Journal of Research*, vol. 1, Nov. 1952.

The Vedic Age, ed. R. C. Majumdar, *History and culture of the Indian people*, vol. 1, Bombay, Bharatiya Vidya Bhavan, 1971.

Velankar, H. D., 'Ṛta and satya in Ṛgveda', S.P. 20th All India Oriental Conference, Bhubaneshwar, 1959, (AIOC).

Venkatesvaran, C. S., 'Vedic conception of asura', *Poona Orientalist*, 13, 1948.

Wallis, H. W., *Cosmology of the Rigveda*, London, Williams & Norgate, 1887.

Werner, K., 'Religious practice and yoga in the time of the Vedas, Upaniṣads and early Buddhism', *Bhandarkar Oriental Research Institute Annals*, Poona, vol. 56, 1975 (BOR Institute).

Werner, K., 'Symbolism in the Vedas and its conceptualisation',*Numen*, vol. 24, farc. 3, 1977.

Werner, K. 'The Vedic concept of human personality and its destiny', *Journal of Indian philosophy*, 5, 1978.

Winternitz, M., *A history of Indian literature*, Calcutta University, 1927, authorised translation into English by Mrs S. Ketkar.

Yamunacharya, M., 'The deeper meaning of yajña in Indian religious thought', *Quarterly Journal of the Mythic Society, Bangalore*, vol. 39, Oct. 1948.

Yevtic, P., *Karma and reincarnation in Hindu religion and philosophy*, London, Luzac 1927.

Zaehner, R. C., *Hinduism*, Oxford University Press, 1966.

Reference books

Bloomfield, M., *Vedic concordance*, Cambridge, Mass., 1906; reprinted Delhi, 1964.

Dandekar, R. N., *Vedic bibliography*, Bombay, 1946; Poona, 1961, 1973.

Grassmann, H., *Worterbuch zum Rig-veda*, Wiesbaden, 1955.

Macdonell, A. A., *Vedic grammar*, 1910; *Vedic grammar for students*, Oxford, 1966.

Macdonell, A. A. and Keith, A. B., *Vedic index of names and subjects*, Delhi, 1967.

Monier-Williams, *Sanskrit-English dictionary*, Oxford, 1970 (reprint of 1899).

Renou, L., *Bibliographie védique*, Paris, 1931.

Word and name index

Index to main texts discussed

Index to main texts discussed

Index to main texts discussed